MATHEMATICAL
MODELING

No. 4

Edited by
William F. Lucas, Claremont Graduate School
Maynard Thompson, Indiana University

John Casti   Anders Karlqvist
Editors

Newton to Aristotle

Toward a Theory of Models for Living Systems

Birkhäuser
Boston · Basel · Berlin

John Casti
Institute for Econometrics and
    Operations Research and
    System Theory
Technical University of Vienna
Argentinierstrasse 8/119
A-1040 Vienna
Austria

Anders Karlqvist
The Royal Swedish Academy
    of Sciences
S-10405 Stockholm
Sweden

ISBN-13: 978-1-4684-0555-2   e-ISBN-13: 978-1-4684-0553-8
DOI: 10.1007/978-1-4684-0553-8

Library of Congress Cataloging-in-Publication Data
Newton to Aristotle : toward a theory of models for living systems /
    John Casti, Anders Karlqvist, editors.
        p.   cm. — (Mathematical modeling ; no. 4)
    Includes index.
    1. Biology—Mathematical models.   2. Biology—Philosophy.
I. Casti, J. L.   II. Karlqvist, Anders.   III. Series: Mathematical
modeling (Boston, Mass.) ; no. 4.
QH323.5.N49   1989
574'.01'5188—dc20   89-7247

Printed on acid-free paper

© Birkhäuser Boston, 1989
Softcover reprint of the hardcover 1st edition 1989

Camera-ready copy prepared by the authors.

9 8 7 6 5 4 3 2 1

Preface

Beginning in 1983, the Swedish Council for Planning and Coordination of Research has organized an annual workshop devoted to some aspect of the behavior and modeling of complex systems. These workshops have been held at the Abisko Research Station of the Swedish Academy of Sciences, a remote location far above the Arctic Circle in northern Sweden. During the period of the midnight sun, from May 4–8, 1987 this exotic venue served as the gathering place for a small group of scientists, scholars, and other connoisseurs of the unknown to ponder the problem of how to model "living systems," a term singling out those systems whose principal components are living agents.

The 1987 Abisko Workshop focused primarily upon the general system-theoretic concepts of process, function, and form. In particular, a main theme of the Workshop was to examine how these concepts are actually realized in biological, economic, and linguistic situations. As the Workshop unfolded, it became increasingly evident that the central concern of the participants was directed to the matter of how those quintessential aspects of living systems—metabolism, self-repair, and replication—might be brought into contact with the long-established modeling paradigms employed in physics, chemistry, and engineering. In short, the question before the house was: Is the world view we have inherited from Newton adequate to understand and formally represent living processes?

Rather early on in the Abisko deliberations, the evidence mounted that something new must be added to the theoretical modeling framework of Newton to account for the peculiar features distinguishing living from nonliving systems. As every college freshman knows, the conceptual framework underlying the Newtonian view of the world is founded upon the twin pillars of particles and forces. This foundation is by now so much a part of the taken-for-granted reality of modern science that it's seldom questioned. Nonetheless, the Abisko participants felt that any kind of "neo-Newtonian paradigm" suitable for living systems will require its own conceptual scaffolding upon which to drape an array of mathematical ideas and techniques for representing the essence of processes in the life, social, and behavioral sciences. As a collectively emergent phenomena, the skeleton of such a conceptual framework arose out of the daily discussions at Abisko. Surprisingly, the consensus view at Abisko was that what is called for is a return to, or more properly, a reconsideration of the world view that Newton overthrew—the world of Aristotle.

Until Newton came along with his ideas of particles and forces, the prevailing epistemology for why events appear as they do was the explanation offered by Aristotle's theory of causes. These Aristotelian causes are four in number—material, formal, efficient, and final causation—and, taken together, they provide a collectively exhaustive and mutually exclusive account for the 'why' of the world. As the contributions to this volume show, a reexamination of these causes through the eyes of modern science and mathematics provides strong hints as to how we might go about constructing a "theory of models" that would play the same role for living systems that the classical Newtonian paradigm plays for lifeless systems. This backward look in time from the conceptual scheme of Newton to that of Aristotle accounts for the title of our volume.

In light of the extremely stimulating presentations and discussions at the meeting itself, each participant was asked to prepare a formal written version of his view of the meeting's theme. The book you now hold contains those views, and can thus be seen as the distilled essence of the meeting itself. Regrettably, one of the meeting participants, Stephen Wolfram, was unable to prepare a written contribution of his very provocative views due to the pressure of other commitments. However, as compensation we have the outstanding contribution by Michael Conrad, a 1986 Abisko "alumnus," who has kindly provided us with a chapter striking to the very heart of the meeting's theme, written moreover in the "Abisko spirit" that he knows so well.

It is a pleasure for us to acknowledge the generous support, both intellectual and financial, from the Swedish Council for Planning and Coordination of Research (FRN). In particular, the firmly-held belief in the value of such theoretical speculations on the part of FRN Secretary General, Professor Hans Landberg, has been a continuing source of encouragement. Finally, special thanks are due to Mats-Olof Olsson of the Center for Regional Science Research (CERUM) at the University of Umeå for his unparalleled skill in attending to the myriad administrative and organizational details that such meetings inevitably generate.

January 1989

John Casti, Vienna
Anders Karlqvist, Stockholm

Contents

Contributors

András Brody—Institute of Economics, Hungarian Academy of Sciences, Box 262, H–1502 Budapest, Hungary

John Casti—Institute of Econometrics, Operations Research, and System Theory, Technical University of Vienna, Argentinierstrasse 8, A–1040 Vienna, Austria

Michael Conrad—Department of Computer Science, Wayne State University, Detroit, MI 48202, USA

David Lightfoot—Linguistics Program, University of Maryland, College Park, MD 20742, USA

Robert Rosen—Department of Physiology and Biophysics, Dalhousie University, Halifax, Nova Scotia B3H 4H7, Canada

Gerald Silverberg—Maastricht Economic Research Institute on Innovation and Technology, Box 616, 6200 MD Maastricht, The Netherlands

René Thom—Institut des Hautes Études Scientifiques, 35 Route de Chartres, 91440 Bures-Sur-Yvette, France

Jan Willems—Department of Mathematics, University of Groningen, Box 800, 9700 AV Groningen, The Netherlands

Introduction

JOHN CASTI AND ANDERS KARLQVIST

1. Process, Purpose, Function, and Form

Reduced to its rock-bottom essence, the goal of theoretical science is to answer the question: "Why do we see the events we do and not see something else"? Of course, the answer to any question beginning with 'Why' starts with the word 'Because,' leading us to conclude that the concern of theoretical science is with *explanations*. And so it is. The theoretician's job is somehow to offer a logical chain of causes that starts with a collection of "primitives" and ends with the observed event to be explained. The epistemological fireworks begin when it comes to specifying just what it is that counts as a primitive.

For the better part of two millenia, the ideas of Aristotle dictated the primitives from which scientific explanations were to be composed. In his theory of causal categories, Aristotle answered the 'Why' question with four mutually exclusive and collectively exhaustive 'Becauses.' According to Aristotle, the events we observe can be explained by their *material, efficient, formal,* and/or *final* cause. To fix this crucial idea, consider the house you live in, an example, incidentally, originally used by Aristotle himself. According to the theory of causal categories, your house takes the form it does for the following reasons: (i) Because of the materials out of which it is constructed (material cause); (ii) because of the energy expended by the workmen who built it (efficient cause); (iii) because of the architectural plan employed in its construction (formal cause); (iv) because of your wish to have a dwelling to protect you from the elements (final cause). Thus, by this scheme there are several ways of answering the question, "Why is my house the way it is"? Interestingly, when wearing his scientific hat, Aristotle was primarily a biologist. Consequently, he attached great significance to living forms and very likely created his theory of causes to explain why living systems appear as they do. In this regard, it's of considerable significance to note that Aristotle reserved his highest regard for final causation, presumably a reflection of the seeming purposeful behavior of most life forms.

About three centuries ago, in one of the greatest intellectual revolutions of all time, Isaac Newton pushed Aristotle's causal explanatory scheme off the center stage of science, replacing it with a radically different way of saying 'Because.' In Newton's world, material particles and forces imposed upon them are the stuff of which events are made. Newtonian reality assumes that the events we observe are formed out of systems of material objects, which are themselves composed of elementary particles. The behavior of these objects is then dictated by forces impressed upon the objects from outside the system. As to the nature of both the particles and the mysterious forces, Newton, cagey as ever, evades the issue entirely with his famous remark *hypothesis non fingo* (I make no hypotheses). With some justice, it might be said that the attempt to address this evasion has provided a good livelihood for physicists ever since.

From an epistemological standpoint, it's of considerable interest to try to relate Newton's world of particles and forces to Aristotle's universe of causes. A little reflection enables us to rather easily match up three of the four Aristotelian causal categories with the main components of Newton's scheme:

$$\text{particles} \leftrightarrow \text{material cause}$$

$$\text{forces} \leftrightarrow \text{efficient cause}$$

$$\text{context} \leftrightarrow \text{formal cause}$$

Here by "context" we mean the background, or environment, against which the particles and forces operate. Thus, things like the gravitational constant, particle masses, electric charges, and so on are part of the context. What's conspicuous about the foregoing match up is the absence of any Newtonian correspondent to Aristotle's final cause. There appears to be just no room for final causation in Newton's world. This is especially troubling when we recall that such a deep thinker as Aristotle reserved his highest regard and consideration for just this way of saying 'Because.' Yet an equally deep thinker, Newton, says in effect that "I have no need for that hypothesis."

At first glance, it would appear that Newton's way of explaining things is vastly inferior to Aristotle's in a variety of ways. First of all, there is no room for any notion of purpose, will, or desire in the Newtonian framework. Moreover, Newton invokes the twin observational fictions of particles and forces to explain the why of things. Yet this so-called "explanation" merely replaces the Aristotelian categories of material and efficient cause by new words. So why is it that a world view and an epistemology that survived intact for almost two thousand years was overthrown virtually overnight by such a seemingly inferior, or at least no more informative, explanatory mechanism? A large part of the answer is bound up with the idea of a model. In particular, a mathematical model.

2. *Mathematical Models*

The selling point of Newton's scheme of things was that he created a mathematical translation of his world view that could be employed for making predictions. And, luckily, he also chose a set of problems (celestial motion) especially well-suited to the explanatory mechanism he had created. In fact, it's amusing to speculate upon the fate of his methods if Newton had instead chosen to focus his attention upon, say, the workings of the brain rather than the meanderings of the planets. But one can never discount luck as a factor in science, and Newton did indeed direct his mathematical apparatus to the solar system and not to the brain. As a result, the dominant paradigm in theoretical science, and the one to which all fields have aspired for the past 300 years, has been the Newtonian vision of what constitutes the right way of saying 'Because.' Since the idea of a mathematical model lies at the heart of the Newtonian paradigm, it's worth taking a moment to consider this kind of "gadget" in a bit more detail.

Every model, mathematical or otherwise, is a way of representing some aspects of the real world in an abbreviated, or encapsulated, form. Mathematical models translate certain features of a natural system N into the elements of a mathematical system M, with the goal being to mirror whatever is relevant about N in the properties of M. The basic idea is depicted in the figure below.

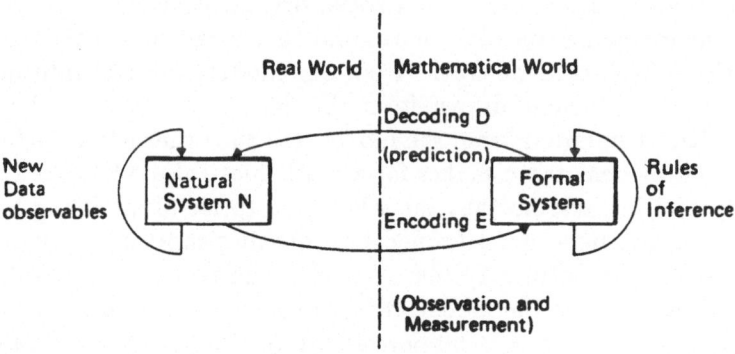

The Modeling Relation

The above diagram shows the two essential aspects of a mathematical model: (i) An *encoding* operation by which the *explanatory* scheme for the real-world system N is translated into the language of the formal system M, and (ii) a *decoding* process whereby the logical inferences in M are translated back into *predictions* about the temporal behavior in

N. So, for example, in Newton's mathematical model of celestial motion, the explanatory principles of particles and forces are encoded (via Newton's 2nd Law) into mathematical objects ($F = md^2x/dt^2$). The mathematical behavior of these objects is then decoded into predictions about the future position of planetary bodies.

For the moment, let's leave aside consideration of exactly *how* the encoding/decoding operations are to be performed. Even with this rather large pile of dirt swept under the rug, a crucial aspect of the success of the modeling process is the selection of exactly which aspects of N are to be encoded into the mathematical system M. In practical situations, this step very often separates success from failure. The crux of the problem usually revolves about what are commonly called "self-evident truths." And it is exactly this sort of truth that is frequently overlooked when we encode the real world into the world of mathematics. A good illustration of this kind of truth is provided by the noted Swedish geographer Torsten Hägerstrand in his consideration of social organization. Hägerstrand notes the fairly obvious facts that a person can be only at one place at a given point in time, and that such a person must be at some spatial location at all times. Despite their self-evident nature, these facts have very definite and often profound implications for how societies are organized. As a result, their omission from the encoding of N into M will have drastic and probably disastrous consequences for the explanation/prediction properties of any mathematical model of a social organization N.

The foregoing considerations point to a paradox. On the one hand, omission of self-evident truths from our models can call into question all of the conclusions drawn from the decoding of the mathematical propositions emerging from the model. On the other hand, inclusion of every self-evident truth makes for a mathematical model that is so unwieldy and intractable that the whole enterprise of modeling becomes self-defeating, being transformed into nothing short of a complete description of the natural system N itself. The way out of this dilemma is to recognize that the model M is, in some definite sense, a compression of the relevant *information* about N into a more maleable and understandable form. From this point of view, "good modeling" reduces to ways to efficiently encode the relevant "truths" about N into mathematical form.

3. Models and Information

In the mid-1960s, the Russian mathematician Andrei Kolmogorov and the Americans Ray Solomonoff and Gregory Chaitin independently suggested a new definition for the complexity of a number sequence. Roughly speaking, their idea was to measure the complexity of a sequence by the length of the shortest computer program required to produce the sequence. Thus, the sequence $0000\ldots000$ consisting of n repetitions of the symbol 0 would not be very complex, since it could always be produced by the program 'Write n copies of the symbol 0.' And this same program would work for any value of n. On the other hand, a sequence like 0110001010111010111001010 appears to have no recognizable pattern, leading to the conjecture that no program appreciably shorter than the sequence itself will suffice for its reproduction. Such a sequence would have high complexity. According to the Kolmogorov-Chaitin view, a sequence requiring a program equal to the length of the sequence itself is the very epitome of randomness. As we'll see in a moment, randomness is the norm with almost every sequence being essentially without any discernible pattern.

A mathematical model can be thought of as a way of encapsulating in a program, or algorithm, the order (structure, pattern) present in a given natural system N. Consequently, we are faced with the problem of trying to compress the information of importance about N into a form in which it can be manipulated by the inferential rules of mathematics. On the other hand, when all the mathematical manipulations are finished, we want to recover the information about N in a usable form, generally a prediction of some sort about the future behavior to be expected from N. This means that ideally the encoding and decoding operations, as well as the mathematical operations on the model, should all be carried out with as little information loss as possible. The degree to which this ideal can be achieved is, in some sense, what separates "good" models from "bad." Unfortunately, creation of "good models" is a very tall order and, in some sense, we should consider ourselves lucky if it can ever be fulfilled.

To see why, let's consider the familiar process of tossing a fair coin. Imagine we code a Head as "1", with a Tail being labeled "0." Then a typical sequence of such tosses might yield the outcome $01100010101010001010111010101\ldots$. This situation is the quintessential example of what we generally think of as a random process. And, in the Kolmogorov-Chaitin scheme of things, almost every such experiment is indeed a purely random sequence of maximal complexity. But now consider the completely deterministic iteration process

$$x_n = 2x_{n-1} \mod 1, \qquad n \geq 1.$$

In this equation, the mod 1 merely means drop the integer part. The equation is thus a mapping of the unit interval onto itself. It's easy to check that its solution is given by

$$x_n = 2^n x_0 \mod 1$$

It is especially revealing to write the initial number x_0 as a binary sequence, e.g., $x_0 = 0.10100111001010\ldots$. Now we can readily verify that the forward iterates of the equation are generated just by moving the decimal point one position to the right and dropping the integer part. It's hard to imagine a more deterministic and easy-to-understand process than this. Yet all the orbits are chaotic and, in fact, are indistinguishable from the coin-tossing situation described above. Let's spend a moment to see why.

Suppose we divide the unit interval into two equal segments and agree that as the iterates of the equation unfold, we will record a "0" if the number is in the left half of the interval and a "1" if the number falls into the right half. When the iteration is complete, the binary sequence obtained from this kind of interval labeling will be identical to the binary expansion of the starting number x_0. Thus, the person marking whether or not the iterates fall into the left or right-halves of the unit interval (the observer) is merely copying down the binary string for x_0. But since we can't in general determine future digits of x_0 from any past finite part of its digit string, the true orbit of the system is chaotic, i.e., unpredictable.

Now consider the situation in which someone with perfect knowledge of the entire orbit of our equation reads out the sequence of digits in x_0. Can we definitively decide if this person is sequentially telling us the first binary digit in each x_n computed from the equation, or is he just obtaining the elements of this digital string by flipping the aforementioned honest coin? It turns out that there's no way to know! For us, the lesson from this example is that for almost every initial x_0 the information contained in the orbit of the system cannot be compressed. Put another way, almost every orbit of our simple equation is of maximal complexity, and the information about the dynamical process cannot be expressed in a program shorter than just reading out the sequence itself. So only in very special situations (in the example, when x_0 is rational, for example) can we ever expect to be able to model a process in a manner much more compact than just letting the process itself unfold.

The same sorts of ideas that apply to process also apply to form, as we can see, for instance, in trying to model the geometric shape of

many natural objects. Mandelbrot's theory of fractals has shown that
the shapes of such irregular geometric objects as snowflakes, coastlines,
and lightning bolts can all be constructed using a self-similar scaling
process of infinite fractal complexity. Thus, while Kolmogorov-Chaitin
complexity shows us the need for what amounts to an infinite amount
of information to represent a dynamical process exactly, Mandelbrot's
theory reflects a different type of infinite information requirement—
geometic forms of infinite extension.

The well-known examples above show that, generally speaking,
to exactly model Aristotle's formal and efficient cause (form and process) involves infinite amounts of information. A similar argument
can be made for material cause, with the high-energy physicist's ever-
expanding list of so-called "elementary particles" a prime candidate as
an exemplar of this unhappy fact. But what about final cause? Here
the situation is less clear cut, not so much because there is any real
reason to doubt the "exactness $\leftrightarrow$ infinite information" coupling, but
more because final cause has for so long been banished from polite scientific discussion. But with the renewed interest today in formalizing
biological processes, there is reason to hope that final cause will again
be accorded equal rights in the community of causes, along with those
more celebrated citizens material, formal, and efficient causation. If
anything, it is the exploration of how we might extend the Newtonian
paradigm to bring about this enfranchisement that is the leitmotiv of
this volume. So without further ado, let's take a brief look at the ways
the authors represented here have addressed these issues.

The book's opening chapter by Robert Rosen goes immediately
to the heart of the modeling relationship depicted in our earlier diagram. Rosen notes that the causal structure associated with a natural
system N is mirrored by the inferential structure associated with a
formal system (mathematical model) M. He then argues that it is an
axiom of modeling faith that the causal and inferential structures can
somehow be brought into harmony with each other. The chapter shows
that this congruence is rather weak in physics, in fact, surprisingly so.
Moreover, Rosen asserts that the Newtonian scheme we have described
above, when turned to problems in biology, requires just the kind of
augmentation associated with the Aristotelian causal view of the world
if it is to meet the needs of living processes. Finally, the chapter gives
some indication as to how this reconciliation might be brought about.

In the second chapter, René Thom continues on the course set by
Rosen claiming that biology will never be a truly theoretical science
until it is able to embed observed phenomena into a larger universe of

"imaginary events" or virtual facts. According to Thom, Aristotle saw clearly the need for virtuality as a necessary condition for theoretical science, formalizing the idea with his distinction between potentiality and actuality. Thom's chapter then shows how this general Aristotelian notion can be employed to classify the physical structure of animal organisms. He concludes by asserting that efficient and final cause can be subsumed under formal cause, at least in biology, by employing the notion of a morphogenetic field.

While both the Rosen and Thom chapters are somewhat general, even philosophical, in character, the third chapter by John Casti tries to show how the Aristotelian causal structure can be mathematically formalized as an extension of the Newtonian paradigm. Using an earlier idea of Rosen's, Casti shows how to bring the crucial functional activities of life—self-repair and replication—into contact with modern mathematical system theory by creating a formal extension of the Newtonian framework. Casti shows explicitly, both by theory and by example, how this new framework works in the case of linear processes, and then indicates a variety of application areas in biology, economics, and industrial manufacturing where the general concepts might be readily employed.

Following up the general theme of modeling, in Chapter Four Jan Willems considers the two central questions of mathematical representations of the real world: (1) Exactly what kind of mathematical objects should we employ in the construction of models, and (2) exactly how should we generate these mathematical models from observed data? Willems addresses these pivotal issues by creating what can only be termed a theory of models, in which notions of model complexity and misfit with the data play central roles. Following a détailed presentation of the arguments underlying his case, Willems concludes by showing that any modeling venture is ultimately a tradeoff between complexity, misfit, and the introduction of auxiliary variables (Thom's "imaginary events").

The first half of the book is devoted primarily to matters of philosophy and modeling theory. Beginning with Chapter Five, the tone shifts to applications in physics, economics, and linguistics. In this chapter, Michael Conrad presents an extensive discussion of two of the central themes in physics—Newton's mysterious forces and the bugaboo of quantum theory, the measurement process. The problem that Conrad addresses is tied up with the conventional view that in all physical processes two distinct types of influences are at work: An influence associated with the forces involved in the exchange of virtual particles

(Thom's "imaginary events" again!), and the influence of the measurement process. The theory presented by Conrad is an attempt to create a modeling picture in which the forces and measurement are both accommodated within the same framework. Conrad's argument is that it is exactly such a modeling paradigm that is needed if the kind of quantum-theoretic setup of physics is to have any chance of making contact with biological phenomena.

Shifting the emphasis from physics to economics, in Chapter Six Gerald Silverberg asserts that from the standpoint of theoretical modeling, economics is still underdeveloped. His argument is that there is no agreement in economics either as to what objects to look at or what basic principles will lead to the identification of appropriate frameworks for analysis. His discussion makes a strong case for moving away from the classical equilbrium-centered view of economic phenomena, instead looking at temporal and structural regularities within populations characterized by diversity and subject to continual evolutionary transformation.

In Chapter Seven, András Brody claims that maybe the distinction Silverberg draws between equilibrium-centered and evolutionary economics is more virtual than real. Brody considers three very different economic world views: (1) The equilibrium-centered view of Adam Smith, (2) the cyclic view of Karl Marx, and (3) the chaotic view of Slutzky. Not surprisingly, these views correspond in one-to-one fashion with the three types of long-term behavior that can be displayed by any dynamical process. Brody shows how it is possible to regard each of these seemingly inimical views as interwoven regimes of a single, simple model of economic growth. He concludes with the observation that the human economy seems to grow in fits and jumps, in a haphazard, fluctuating, but neverthelesss relentless manner.

The book's final chapter centers upon that most interesting of all living systems—a human being. In particular, Chapter Eight by David Lightfoot focuses upon the unique human trait of spoken language. Vigorously pressing the claim that human language acquisition and development is dictated by genetic programs, Lightfoot offers a model of linguistic development that might explain how language systems change from generation to generation. With this chapter, the book comes full circle back to a causal explanation of a living system that seems totally incomprehensible when viewed from the vantage point of Newtonian physics. Yet when looked at through Aristotle's eye, Lightfoot's arguments seem perfectly consistent with an explanation along the lines of causal categories, lacking only the kind of formal structure that in principle might be supplied by the theoretical machinery developed in some of the book's earlier chapters.

On balance, the inescapable conclusion that emerges from the all-too-brief Abisko excursion into the world of Aristotle is that there really is something different about living systems. So in order to have modern ideas on modeling make contact with this "something different," we're going to have to seriously reconsider the paradigmatic framework within which we spin our models of reality. If nothing else, the contributors to this volume have given us a well-filled plate of *hor d'ouevres* to start the banquet!

The Roles of Necessity in Biology

ROBERT ROSEN

Abstract

Any system is characterized by the entailments mandated within it. In formal systems, these entailments take the form of inferences governed by explicit production rules. In natural systems, entailments are governed by causality. It is an article of faith in science that the two modes of entailment can be brought into a congruence in such a way that inference in a formalism mirrors causality in the world, and conversely. This congruence is explicitly embodied in a modelling relation between a natural system and a formalism

We investigate the entailment structure characteristic of modern physics, which we argue is surprisingly weak. In fact, it manifests itself entirely in a recursive sequence of state transitions, which is itself determined by things unentailed within the formalism. This weakness in entailment makes the formalism appear very general, in terms of what can be encoded into it, but makes the formalism very special as a formalism. We contrast it with the entailments required in biology on the one hand, and with the much broader province of causality originally envisioned by Aristotle, and argue that (a) biology requires modes of entailment not presently available in any physical formalism, and (b) the old Aristotelian view of causality is far more consonant with the exigencies of biology. We indicate a way in which these several observations might be consistently reconciled.

1. Introduction

" ... Life can be understood in terms of the laws that govern and the phenomena that characterize the inanimate, physical universe ... at its essence, life can be understood only in the language of chemistry.

... Indeed, only two major questions remain enshrouded in a cloak of not quite fathomable mystery: the origin of life ... and the mind-body problem ... "

These sanguine words were written by Philip Handler, in his preface to the book *Biology and the Future of Man.* This book was a comprehensive survey of biology as it was in 1970, and as it essentially remains today. It was a paean to Molecular Biology and to the powers of Reductionism.

In what follows, we are going to concentrate on the three words "not quite fathomable." The grudging "not quite" constitutes an admission that the problems addressed are hard, but suggests that the difficulty only resides in our being insufficiently fluent in the language of chemistry. Handler does not admit that the ultimate questions about

life are written in some other language, a language which translates
only imperfectly, or even not at all, into chemistry and physics as we
now know them. If this is so, we must allow that it is the language
of chemistry, and ultimately the language of physics on which it rests,
that must be translated into this new language, to the extent this is
even possible, in order to make these questions fathomable, or even
intelligently articulable.

Let me give another quotation to illustrate the kind of thing I
mean. Many years ago, Edgar Allan Poe described the search for a
Purloined Letter, a problem also "not quite fathomable" to the police who were searching for it. Poe's detective, Dupin, describes the
situation this way:

> "The Parisian police are exceedingly able in their way. They are
> persevering, ingenious, cunning, and thoroughly versed in the knowledge
> which their duties seem chiefly to demand ... Had the letter been de-
> posited within the range of their search, these fellows would, beyond a
> question, have found it.
> The measures, then ... were good in their kind, and well executed;
> their defect lay in their being inapplicable to the case ... [The police]
> consider only their own ideas of ingenuity; and in searching for anything
> hidden, advert only to the modes in which *they* would have hidden it ...
> They have no variation of principle in their investigations; at best, when
> urged by some unusual emergency—by some extraordinary reward—they
> extend or exaggerate their old modes of *practice*, without touching their
> principles. What is all this boring, and probing, and sounding, and
> scrutinizing with the microscope ... what is it all but an exaggeration
> of one set of principles of search ... ? You will now understand what I
> meant in suggesting that, had the purloined letter been hidden anywhere
> within the limits of the Prefect's investigation—in other words, had the
> principle of its concealment been comprehended within the principles of
> the Prefect—its discovery would have been a matter altogether beyond
> question."

Biology is harder than the search for the Purloined Letter, in large
part because the position espoused by Philip Handler makes it an essential part that there are no other methods of search besides "boring, and
probing, and sounding, and scrutinizing with the microscope." Indeed,
according to a still more articulate postulant of this position (Jacques
Monod), the assertion that there are other principles of search (or,
what is the same thing, that there is a new physics to be learned from
a study of organisms) is vitalism, and thus beyond the pale.

Although it is not unusual for a theology to claim that it is already,
at some particular time, all-encompassing and universal, it is most unusual for a science to make such a claim. Therefore, it is instructive
to look briefly at the epistemological presuppositions underlying this

assertion; they are interesting in themselves, and will turn out to lead naturally into our main subject-matter, to be outlined below.

It was Descartes who initially proposed the "Machine Metaphor," which provides one essential prop for modern Molecular Biology. Apparently, as a young man, Descartes was much impressed by some lifelike hydraulic automata he had seen in the gardens of some chateau. The superficial similarities between the behaviors of these automata, and the behaviors exhibited by organisms, led him ultimately to assert, not that machines can sometimes exhibit lifelike behavior, but rather that lifelike behavior is always the product of an underlying machine. In other words, organisms form a proper subclass of the class of machines, and the study of biology is subsumed under the study of machines or mechanisms.

This breathtaking assertion provided, as it turned out, a way of studying biology without seeming to invoke any of the murky concepts associated with Aristotelian Finalism, of which we will see more later. But it was undertaken with only the most shaky conception of what a machine is, and an even more rudimentary conception of what is an organism.

A generation or two later, Newton provided an indirect answer to the question "what is a machine?" in his creation of the science of particle mechanics. At root, Newton turned back to the views of the pre-Socratic Greek atomists; namely, that all substance could be reduced or resolved into ultimate, structureless atoms, possessing nothing inside them which could change in time, and consequently possessing no attributes but position or configuration (and thus also what we would now call the temporal derivatives of configuration). All material phenomena thus devolve upon the particulars of the motions of constituent particles, as they are pushed around by the forces impinging upon them. This, as we shall stress later, is a very syntactic view of the material world, but one which remains compelling in physics itself. In any event, insofar as any material system could be subdivided into constituent particles, the Newtonian theory provided the basis for a truly universal theory of material nature.

This provides the second basic prop on which Molecular Biology rests. For it basically asserts that *every* material system is a machine or mechanism; insofar as any material system inherits its gross behaviors from the motions of its constituent particles, and insofar as the motions of these particles are themselves mechanical in nature, there are in fact no material systems which are not machines in this sense. Thus we have the following inclusions which constitute the Trinity of Molecular Biology:

$$\text{organisms} \subset \text{machines} \subset \text{mechanisms}$$

It is instructive to see what has become of Newtonian particle mechanics within physics itself. Newton, in his search for "universal laws," clearly believed that the same laws manifested in the behavior of ordinary objects must also hold good at every other level, from the atoms themselves to the galaxies. This absolutely basic assumption, which passed unquestioned (and even unarticulated) for three centuries, turned out to be completely false. For instance, not only did it turn out that real, physical atoms do possess internal stucture (contrary to hypothesis) but it turned out further that the laws governing this internal structure are quite different from the Newtonian. These same laws likewise fail, for quite different reasons, when we turn to astronomical scales. Indeed, in retrospect, it is almost miraculous that physics as a science could survive such a lethal invalidation of its most basic hypotheses with as little damage as it has, but that is another story.

Returning now to the inclusion of organisms within the category of automata, and of automata within the category of mechanisms, we see that its immediate effect is to obliterate any distinction between the organic and the inorganic. This is in fact the basis for the reductionistic assertion that biology will be subsumed under physics, by which is meant that very same physics which sits on the right-hand side of our chain of inclusion. It further follows that the way to properly study an organism is the way appropriate for the study of any material system, organism or not; namely, find and isolate the relevant constituent particles, describe how they move under the action of impinging forces, and extract from this ultimate information that bearing on the behaviors of initial interest.

This is Reductionism. It is a happy theory for experimentalists, for several reasons. First of all, it seems to leave the theoretician nothing further to do; no way to meddle further in the ongoing business of science. On the other hand, it gives the experimentalist plenty to do; the isolation and characterization of the relevant constituent particles is obviously an empirical job. True, dimly lurking on the horizon, are the "not quite fathomable" mysteries at the core of biology. But it is easy to ignore these, or to rationalize them with the words of Poe's Parisian Prefect of Police: "The problem is **so** simple."

The belief in Reductionism, which we have sketched above, brings the concept of **Necessity**, or as we shall prefer to say of **Entailment**, into the picture for the first time. As it appears in Reductionism, it is the assertion that all behaviors of organisms are **Entailed** by the Laws of Mechanism. Stated another way: A study of matter through reduction to constituent particles loses no shred of information pertaining to the organization, to the *life* of the system under study. It remains there, though perhaps a little transformed, a little hidden, but always there; and if there remain questions "not quite fathomable," the difficulties arise not from want of information, but only from insufficient cleverness in extracting it. In other words, what difficulties there are are of a *logical* character, which prevents us from making the postulated entailments manifest.

On the other hand, this mechanical picture has always run into trouble because it seems to *entail too much* about organisms. Indeed, those properties of organisms that are most immediate, most conspicuous, seem immune to the kinds of mechanical entailment that rules inorganics with such an iron hand. To this, Molecular Biology provides several stock answers: (1) The apparent freedom from entailment manifested by organisms is an illusion; in reality they are executing fixed *programs*, generated through evolution by Natural Selection. There has even been a new word coined to describe this process: *teleonomy*. (2) On the other hand, there are crucial biological processes which are actually exempt from necessity; exempt, indeed, from entailment of any kind. Evolution itself is such a process. As a result, the evolutionary process cannot in principle be predicted; it can only be *chronicled*. This view has the advantage of exempting the most important parts of biology from science altogether; biology thus becomes a part of *history* and not of science (where by history we mean precisely *chronicle without entailment*). That these two answers are contradictory does not seem to trouble anyone very much.

We are going to suggest in what follows that the "not quite fathomable" problems at the heart of biology arise because all of the principles of search which we have enunciated above are wrong. The inclusion of organisms in the class of automata, which we owe to Descartes, is wrong; the idea that all material systems are mechanisms is also wrong. The very idea of entailment, or necessity, which we inherit from these traditions, is inadequate to deal with the phenomena of life. Indeed, we shall end up by arguing that the inherited inclusions of organisms within automata within mechanisms goes more the other way.

2. Generalities Regarding Entailment

There are two parallel realms in which the concept of necessity or entailment manifests itself. One of these is the realm of the external world, of the processes of nature. We have stressed before that one must believe that the sequences of events which we perceive as unfolding in the external world are not entirely arbitrary or whimsical, but rather manifest general relations, one to another. If so, this lack of arbitrariness is expressed in the form of relations between events in unfolding sequences; such relations are generally referred to as *causal*. Thus, causality in general is the study of entailment or necessity as it is manifested in the external world of phenomena; i.e., it is the subject matter of the sciences.

On the other hand, there is also the internal realm of ideas, which in the broadest sense is the realm of language, symbol, and formalism. This world is not populated by phenomena in the usual sense, but by propositions, which have a different kind of existence from phenomena, and are relatable to the latter only in obscure ways (of which more later). But just as events in unfolding sequences are related, via causality, so too are propositions. One fundamental kind of relation between propositions, which in many ways parallels the causal relation between phenomena, is that of *implication* or *inference*. The study of this relation is the study of necessity or entailment in the internal symbolic world.

We have asserted elsewhere (e.g., Rosen, 1985a) that Natural Law consists essentially of belief that the two great realms of entailment or necessity can be brought into some kind of congruence. In particular, it consists of the belief that causal sequences in the world of phenomena can be faithfully imaged by implications in the formal world of propositions describing these phenomena. The exact statement of this belief is encapsulated in a kind of commutative diagram shown in Fig. 1 expressing a modelling relation between a class of phenomena (i.e., a natural system) and a formalism describing this class of phenomena:

Here, commutativity means explicitly that

$$\text{arrow } 1 = \text{arrows } 2 + 3 + 4$$

i.e., we get the same answer whether we simply watch the sequence of events unfolding in the external world (the arrow 1), or whether we encode into our formalism (the arrow 2), employ its inferential structure to prove theorems (the arrow 3), and decode these theorems to make predictions about events in the external world (the arrow 4). If commutativity holds, we can then say that our formalism is a *model* of the phenomena occuring in the external world, or equivalently that the

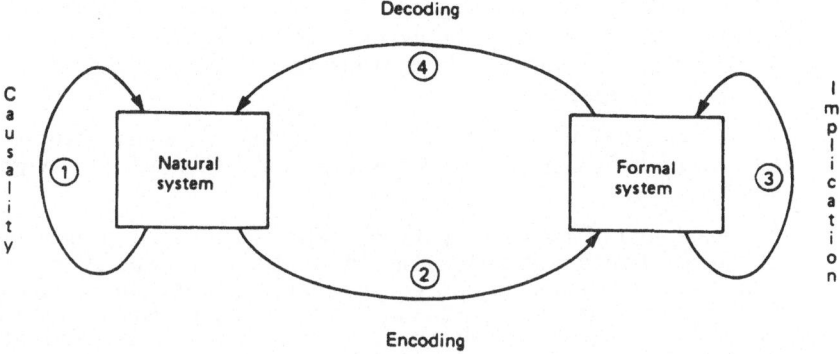

Figure 1

events themselves constitute a *realization* of the formalism. For fuller details regarding the ramifications of these ideas, see Rosen, *loc. cit.*

We point out here, for future reference, that the arrows 2 and 4, which we labelled "encoding" and "decoding," are not themselves entailed by anything, or at least not by anything present in either the formalism (the model) or the external realization of that formalism. Moreover, as we shall see abundantly in a moment, there are also many aspects of formalisms (and hence of their realizations) which are likewise not entailed. Indeed, an essential part of the discussion to follow has to do with the escape from entailment and the intimate involvement of the basic questions of biology with this escape.

Accordingly, in the next few sections, we shall be concerned with a closer analysis of the notion of necessity, or entailment, in the two great realms of phenomena and formalisms, and the relations between them. Armed with this analysis, we shall then investigate their significance for both.

3. Causality: Necessity In the External World

Historically, the Mechanics of Newton manifested a striking break with everything which had gone before. With him, the concept of causality became a very different thing than it had been previously. And in the present century, with the advent of a new mechanics (quantum mechanics), it has apparently changed radically yet again, to the point where no two physicists can now agree on what it means. Consider, for example, the following few quotations on the subject (which could easily be multiplied manyfold):

1. "The necessary relationships (in the sense that they could not be otherwise) between objects, events, conditions, or other things at a given time and those at later times are ... termed causal laws." (D. Bohm, 1957)

2. "The fact that initial conditions and laws of nature completely determine behavior is ... true in any causal theory." (E. Wigner, 1967)

3. "Consistency of nature may be characterized by saying: as a result of the constitution of nature, the differential equations by means of which it is described do not contain explicit functions of time ... Consistency, the central issue of the causality postulate, banishes absolute time from the descriptions of nature ... by eliminating time explicitly from its *fundamental* representations." (H. Margenau, 1950)

4. "No property at time *t* is determined—or even affected—by the events that may occur thereafter." (B. d'Espagnat, 1976)

5. "The assumption underlying the ideal of causality [is] that the behavior of a physical object ... is uniquely determined, quite independently of whether it is observed or not ... the renunciation of the idea of causality ... is founded logically only on our not being able any longer to speak of the autonomous behavior of a physical object ... " (N. Bohr, 1937).

Clearly there is no consensus here; indeed, it is far from clear in what sense these authors are even talking about the same thing. Nevertheless, the grim, persistent attempts to come to terms with causality make it clear that the concept remains essential to the basic enterprise of physics.

To try to clarify the situation, let us return once again to Newtonian mechanics, where in some sense the trouble started. This is doubly important, because Newtonian mechanics has imparted its form, and all the presuppositions embodied in it, to every subsequent mode of system description known to me.

At the very first step—so trivial that it was never even noticed explicitly—the Newtonian analysis made an essential dichotomy between *system* and *non-system*, i.e., between system and everything else, between system and environment. In Newtonian particle mechanics, this distinction is absolute, once-and-for-all. *System* means some definite family of particles to be followed forever over time; environment is whatever else there may be in the world. Subsequently, the two receive entirely different treatments; entirely different representations in the Newtonian image of the world. What is system, for instance, is described by phases or states; environment is not, and cannot, be represented in such terms. Rather, environment is the seat of (external) *forces,* manifested in the equations of motion which is imposed on the states or phases that describe system. Environment is, further, the seat of whatever it is that sets initial conditions, initial configurations, and initial velocities.

This apparently necessary and innocent partition of the world into system and environment, with the resulting difference in description and representation accorded to the two, has had the most profound consequences for the notion of causality. For according to it, the realm of causality becomes bound irrevocably to what happens in system alone; and what happens in system alone is the state-transition sequence. We cannot even talk about environment in such terms; what happens in environment has thus been put beyond the reach of causality. Environment has become *acausal.*

It is true, of course, that we can always reach into this acausal environment and pull another *system* out of it, thereby bringing another part of environment under the provenance of causality once more. But to suppose that *the whole universe* can be described as one big system, with an empty environment—as, for example, Laplace supposed—is quite another matter, involving a totally new supposition about the world. What is anyway clear is that, as long as our system is circumscribed in any way, there must be sitting outside of it an acausal environment, in which *nothing is entailed* in the conventional sense— because there is nothing in its description which even allows a concept of entailment in the first place.

In fact, the traditional domain of conventional physics—closed systems, conservative systems, even dissipative systems—involve *hypotheses* about the environment, which are of their very nature unverifiable. For, in Newtonian terms, environment is created by the same act as that which created system. What happens in it is not entailed, as we have seen, and is hence entirely unpredictable. To say that we *can* predict it, as we assert when, for instance we say a system is closed or conservative or isolated, is precisely one of those hypotheses which Newton was so proud of never having to make.

Thus, as we have said, the essence of the Newtonian picture is precisely to constrict the realm of causality, and hence of necessity, to the state-transition sequence in *system.* This is in fact the sole unifying thread in all the quotations about causality with which we opened this section. Entailment in this picure is *entailment of next state from present state,* under the influence of the acausal external world, according to the rule

$$(x_o, v_o) \rightarrow (x_o + v_o \, dt, \, v_o + F \, dt).$$

And *everything* appearing in this statement of entailment—the initial state x_o, v_o, the force F, and even the time-differential dt—all this is *not entailed*, with its seat in the unknown, undescribed, acausal environment.

In quantum mechanics, the situation is a little different; in some ways better, but in some ways worse. The basic problem is that in classical Newtonian physics state (the seat of causality) is defined in terms of observations which are not allowed in quantum theory. Thus, if we want to retain the causality of the state-transition sequence, we must redefine the concept of state. This is indeed what is done; the concept of state is defined in quantum mechanics in precisely such a way that causal state-transition sequences are retained, but the relation of that state to the Newtonian one has been given up. But the decisive partition of the world into causal system and acausal environment, the essential feature of the Newtonian analysis, is still there in quantum theory; perhaps even more troublesome because it is even more restrictive.

Let us turn now to what it was that the Newtonian picture replaced. This, of course, means essentially Aristotle. In Aristotle, as we shall see, the notion of entailment, of necessity in nature, was far wider than in the Newtonian paradigm which supplanted him. In Aristotle's view, science itself was to be concerned with what he called "the why of things." By "things," Aristotle apparently meant something far more embracing than that embodied in the concept of state; he meant not only a part of an event, or a whole event, or a sequence of events, but *systems themselves.* As we shall see, this latter is something we cannot even legitimately frame in the Newtonian context, but it comes to be of the essence in biology.

The answer to a "why?" is a "because." Aristotle suggested that there are precisely four different, inequivalent, but equally correct ways to say "because"; each one necessary, and all together sufficient, to understand the thing. These were, of course, his Categories of Causation: material cause, formal cause, efficient cause, and final cause or *telos*. Each of these Categories of Causation, in its own way, necessitated or entailed the thing itself; the thing thus became the *effect* of its causes. The establishment of causal relationships from cause to effect created *chronicles,* but unlike purely historical chronicles these were governed *entirely* by relations of entailment. It was the business of science to construct such chronicles.

As we have argued at great length elsewhere, the ghosts of three of the four Aristotelian categories of causation remain in the Netwonian paradigm (cf. Rosen, 1985b). But the only *chronicle* that remains is, as we have said, that of state-transition sequence within a system. And more than anything else, it is the fact that no room remains within the Newtonian paradigm for the category of Final Causation which is responsible for banishing *telos* from any place in modern science.

We nowadays try to do biology in terms of the notions of causality, or entailment, which we have inherited from Netwon, not from Aristotle. We seek to discover the nook in which the key to biology is hidden; we seek to answer the question "what entails biology?" within that framework. As we shall argue subsequently, there is simply not enough entailment left in the Newtonian picture to even frame this question, let alone answer it. Consequently, there is no such nook. That is the ultimate reason why the basic questions of life remain "not quite fathomable."

4. Necessity in Formal Systems: Inference and Implication

Just as Newton provided the bellwether for modern ideas about entailment in the external world, so did Euclid provide the basic model for entailment in the internal world. Indeed, Euclid provided the first, and for many centuries the only, real example of what we would today call a formal system, or formalism. And as we shall see, the language of states and dynamical laws which characterize the Newtonian picture find obvious counterparts in the formal notions of propositions and axioms (production rules), which produce new propositions from given ones.

The fundamental Euclidean picture of a formal system is as a set of statements or propositions, all derived from a few initially given ones (postulates) by the successive application of a number of axioms. This embodies the notion of impliction in the formalism, and with it the notion of *logical* necessity; logical entailment. Insofar as we regard the postulates as *true*, the laws of implication in the formalism propagate this truth hereditarily from postulates to theorems, and from theorems to theorems.

Nowadays, however, we would regard this Euclidean system as "informal." Indeed, the word "truth" as we have used it above, would be excluded. For to say that a propostion is true is to say that the proposition is *about* something; specifically, about some external referent outside the system itself. Euclid, for example, manifestly took it for granted that the propositions in his *Elements* were about geometry; that at least some of the truth in his system arose precisely from this fact.

Let us put the issue another way. In any linguistic system, there are some truths which are purely a matter of form; they arise simply by virtue of the way in which the language is put together. Such truths are independent of what is asserted by a proposition in the language, and depends only on the *form* of the proposition. We shall call such a truth a *syntactic,* or formal, truth. There are, however, other truths, which do depend on what is asserted by the proposition. For want of a better word, we shall call such truth *semantic* truth, where by "semantic" we understand only "non-syntactic." Every language, including Euclid's *Elements,* includes inferential rules governing both kinds of truths.

However, the syntactic kinds of truth in a language seem somehow more objective than those which depend on meaning or signification. Particularly in mathematics, it has seemed that the more syntactic truth, and the less semantic truth, which a language possesses, the better that language must be. In the limit, then, the best language would be one in which *every* truth was syntactic. And in our century, this has been perhaps the main goal of mathematical theory, if not of practice; to replace all "informal," semantic inferential processes with *equivalent* syntactic ones. This process, for obvious reasons, is called *formalization* and has never been better described than by Kleene (1951):

> "This process [formalization] will not be finished until all of the properties of the undefined or technical terms of the theory which matter for the deduction of theorems have been expressed by axioms. Then it should be possible to perform the deductions treating the technical terms as words in themselves without meaning. For to say that they have meaning necessary to the deduction of the theorems, other than that which they derive from the axioms which govern them, amounts to saying that not all of their properties which matter for the deductions have been expressed by axioms. When the meanings of the technical terms are thus left out of account, we have arrived at the standpoint of formal axiomatics ... Since we have abstracted entirely from the content matter, leaving only the form, we say that the original theory has been *formalized.* In this structure, the theory is no longer a system of meaningful propositions, but one of sentences as sequences of words, which are in turn sequences of letters. We say by reference to the form alone which combinations of words are sentences, which sentences are axioms, and which sentences follow as immediate consequences of others."

Clearly, such a formalization of a mathematical theory, such as Euclidean geometry, will not look much like the original theory. But Formalists such as Hilbert clearly believed that any meaningful theory whatever could be dumped into a formalist bucket without really losing any of its "meaning"; i.e., without losing any of the "truth" present in the original informal system. What we intuitively call "meaning" and "truth" are simply transmuted into another form; a purely syntactic form, by means of additional syntactical structure. And if we should for some reason want to re-inject an external referent into the formalism, we can always do so by means of a "model" of the formalization (effectively, by what we earlier called a realization of it).

We raise these issues here for two reasons. First, because of the obvious parallels between what we have above called a formalization of an inferential system, in which all inference is replaced by syntactic inference alone, and the Newtonian particle mechanics, with its structureless ("meaningless") particles pushed around by impinging ("syntactical") forces. As we shall soon see, this bears directly on the "machine metaphor" of the organism, to which we have already made reference above. Specifically, we shall see how this brings Newtonian ideas of entailment to bear directly on the problems of biology. In particular, we can already perhaps see a close parallel between formalization in the formal realm and reductionism in the material. We shall now turn to a second reason, because it will turn out to be of even more importance.

In our discussion of the role of entailment in the natural world of phenomena in the preceding section, we pointed out the essential role of the partition between system and environment. In the Newtonian paradigm, the effect of that partition was to restrict the role of causal necessity entirely to system, and even there, to embody it only in the transition from given state to subsequent state. What needs to be pointed out is that there is an exactly parallel duality between system and environment tacit *in any formal system,* like Euclid's, or any formalization thereof. Just as before, entailment is a concept restricted only to such a system; it is inapplicable in principle to the great sea of other propositions from which system has been extracted. These propositions are *not entailed;* nor are the processes by means of which the system postulates (initial conditions) or axioms (inferential rules or dynamical laws) were pulled out of environment and into system in the first place.

Let us put these basic ideas into a more tangible form. We can assert that the prototypic syntactic inferential process finds its mathematical form in the *evaluation of a mapping,* and thus can always be put into the form

$$a \Rightarrow b = f(a).$$

In words, then, we can always say: *a implies b*, or *a entails b*, according to the inferential rule designated by f.

Actually, it would be more accurate to put this implication into the form

$$f \Rightarrow (a \Rightarrow b = f(a)); \tag{1}$$

that is, the inferential rule f entails that a entails $b = f(a)$. This usage is not just a refinement; it becomes absolutely mandatory when there is more than one inferential rule available.

Let us contemplate the above statement (1). The first thing to notice is that the implication symbol "$\Rightarrow$" is actually being employed in two quite different senses. Namely, if we were to look at $b = f(a)$ as analogous to the Aristotelian notion of "effect," then the relation between $f(a)$ and a itself, expressed in the notation $a \Rightarrow b$, would be analogous to the relation of effect to *material cause;* on the other hand, the relation between f and b, governed by the other implication symbol, is of a quite different type, analogous to the Aristotelian relation of effect to *efficient cause.*

The next important thing to notice is that, in (1), the only entailment in sight is that of $b = f(a)$. Nothing else in this expression is entailed by anything. That is, neither the element a, which can variously be regarded as the "input," or as the "initial condition," nor the operator or inferential rule f, is entailed by anything. Stated another way: neither a, nor f, can itself be regarded as an effect. Intuitively, they are simply pulled in from the vast sea of non-entailed *environment,* by means of unspecified and unspecifiable processes, to comprise the simplest example of what we would call a *system.*

Let us next consider a slightly more general situation. Suppose that our inferential rule is actually defined, not directly on the set A itself, but on a larger set; say

$$f: A \times \Sigma \rightarrow B.$$

Then for each element $\sigma \in \Sigma$, we regain thereby a mapping

$$f_\sigma : A \rightarrow B$$

defined by $f_\sigma(a) = f(a, \sigma)$. If our inferential rule is now taken to be f_σ instead of f, we can ask: what is the relation between σ and the effect $f_\sigma(a)$? Or in other words: in what sense can we write

$$\sigma \Rightarrow f_\sigma(a) \ ?$$

The relation between σ and the effect $f_\sigma(a)$ is obviously analogous to the Aristotelian relation between *formal cause* and effect. Once more, the inferential symbol "$\Rightarrow$" is being used in quite a different sense from before. And obviously, in this process, we introduce yet another entity into our system which is not entailed, the element σ.

It may be helpful to point out here that, in a certain sense, the element we have called σ may be regarded as a *coordinate,* which locates a particular mapping or inferential rule in a larger space of such mappings. Thinking of it in this way clearly demarcates it from either the mapping which resides at that address, or the actual domain on which that mapping operates. Alternatively, σ may be thought of as set of *specific parameter values,* which must be independently specified in order for the mapping f itself to be completely defined. This kind of "independent specification" is what is unentailed, along with the initial conditions $a \in A$, and the operator f itself, each in their different ways.

We will now briefly take up the subtle and difficult discussion of the remaining Aristotelian causal category, that of final causation or *telos,* in this elementary setting. Its central importance to our enterprise will become clear subsequently.

Telos turns out to be related to entailment in the opposite logical direction to the considerations we have just developed. That is, the Aristotelian usage, the "final cause" of something relates to *what is entailed by it;* and *not to what entails it.* Thus, throughout the above discussion, we have treated $f(a)$, or $f_\sigma(a)$, as the effect, entailed in different ways by its causes a, f and σ. But clearly we cannot in the same way speak of a *final* cause for $f_\sigma(a)$, because *there is nothing in the system for it to entail.*

Instead, we could only speak, for instance, of the final cause *of a,* or the final cause *of f_σ,* or the final cause *of σ,* because these in fact *do entail something* in our system. In fact, they all, in their separate ways, entail their effect $f_\sigma(a)$. We could say that their *function* in the system is precisely that of entailing this effect. And to speak of function in this sense is exactly to speak of telos.

It is crucial to notice at this point that telic entailment, or final causation, serves precisely to "entail" (in its way) *everything which is in fact unentailed in the "forward" causal direction.* It does this by interchanging the specification of what is (final) cause, and what is effect in every other causal sense.

We now come to the crux of our discussion. Suppose we do want to entail that which is so far otherwise unentailed in our system. Suppose we do want to entail, for example, the efficient cause f_σ of the effect

$f_\sigma\,(a)$. That is, we now want to be able to treat the operator f_σ itself as an *effect,* and thus to be able to speak of its own material, efficient, and formal causes. The effect of this is to make telic, backward causation, and this new forward causation, *coincide.* This is to say: any mode of entailing a causal agent, in the "forward" causal direction, is equivalent to its telic entailment in the "backward" direction.

Now obviously, if we want to treat an operator like f_σ as itself an effect (i.e., if we want it itself to be entailed by something) we need to *enlarge our system* with additional inferential structure, of a kind to be described below. This additional structure is of a kind normally excluded from formalisms themselves; either purely mathematical ones, or ones which purport to image external reality. This additional structure will turn out to possess almost magical properties; it will at one stroke provide us with a formal basis to discuss *fabrication of systems,* in which whole systems, and not just states of systems, can be treated as effects (i.e., discussed in terms of entailment or necessity), but it will allow us to treat telos in a perfectly respectable way. Telos has been so troublesome precisely because the basis for discussing it has always been left out of the kinds of entailment we have learned to call *system,* and relegated entirely to the structureless, acausal environment.

And conversely, if we do wish to speak about the "final cause" of $f_\sigma\,(a)$, we must add corresponding structure in the other direction, in order to *give* $f_\sigma\,(a)$ *something to entail.* Both these modes of enlargement, as we have argued, take us out of the traditional "system paradigm" as we have inherited it. As we shall see, it is only by leaving this paradigm that we will, in a sense, acquire enough entailment to do biology. But if we do leave it, we will also see that we change our view of physics, and even of formalism itself, in unexpected ways.

Let us then begin to indicate how we can supply what is required. According to the above discussion, what we conventionally refer to as *system* leaves unentailed most of what appears to the "logical left" of an effect or inference: the "initial conditions," the logical operator itself, and whatever paramenters on which the logical operator itself depends. Likewise, it leaves little room for that effect or inference to itself entail anything. We shall take these matters up in turn.

We will focus our attention specifically on the entailment of a logical operator, what we called f, or f_σ, in the course of the above discussion. This operator, it will be recalled, serves as "efficient cause" in the entailment of its "effect" $f(a)$, or $f_\sigma\,(a)$. We thus ask the question: how can we entail this *operator* f? Or equivalently: how can we speak of this operator as itself being an effect of causes?

We can formally approach this question by imitating our analysis of the operator f itself. Namely, we want to be able to write an expression of the form

$$(?) \to f,$$

analogous to our previous statement of entailment,

$$a \Rightarrow f(a).$$

It was, of course, precisely for this purpose that we explicitly introduced the operator f, to enable us to write

$$f \Rightarrow (a \Rightarrow f(a)).$$

Thus, analogously, it is clear that we need another operator, Φ, to enable us to write an entailment of the form

$$\Phi \Rightarrow ((?) \Rightarrow f). \tag{2}$$

This new operator Φ, it will be noticed, is one which *entails another operator,* and thus belongs to a new logical level, different from that on which the operator f itself operates. It accomplishes this entailment by operating on an as yet unspecified substrate which we have called (?); thus making this substrate function as "material cause" *of the operator* f.

So far, we perhaps seem to have gained little. Although we now can formally entail f, we have done so at the expense of introducing another operator Φ, which is itself unentailed. And we have also added a new unentailed substrate (?) on which that operator is defined. Finally, we have still given the original "effect" $f(a)$ nothing to entail. Indeed, all we have accomplished is to recognize explicitly that if the original operator f is itself to be entailed, we require for that purpose a new operator Φ, of a different logical type.

We can formally resolve most of those problems in the following way. Suppose that we formally require

$$(?) \equiv f(a). \tag{3}$$

If we do this, then at one stroke we have: (i) given the original effect, $f(a)$, *something to entail,* and (ii) removed the need for an independent entailment of the substrate (?). Thus, this larger system actually contains *more* entailment (or alternatively, less unentailment) than did the original system with which we started. Indeed, all that is now left unentailed in the larger system governed by the relations (1), (2), (3) above are (i) the original material cause a of the effect $f(a)$, and (ii) the new operator Φ.

However, we can now re-invoke our little trick, embodied in (3) above, of enlarging the original system by effectively identifying forward and backward logical entailment, to see if we cannot find already *within our system* a means of entailing the operator Φ itself. If we could do this, we would at a single stroke obviate the apparent incipient infinite regress of operators to entail operators to entail operators ..., and at the same time, limit the non-entailment remaining within the system to the maximal extent possible, namely, to the original material cause a of the original effect $f(a)$. That is, we will thereby have constricted the role of the (acausal) environment entirely to this factor.

Curiously enough, a formalism with all these properties already exists, although it was developed for completely different purposes. A system consisting of the operators (f, Φ), satisfying the above conditions, is precisely an example of what we earlier called an (M,R)-system (cf. Rosen, 1958); an (M,R)-system in which the operator Φ is itself entailed from within the system is an (M,R)-system with *replication* (cf. Rosen, 1959). The (M,R)-systems were originally intended as a class of relational models of biological cells, with the operator f corresponding to vegetative, metabolic, or cytoplasmic cellular processes, and the operator Φ corresponding to the nuclear or genetic processes. What we earlier called replication becomes, in this context, precisely the entailment of the genetic part of the system (the operator Φ) from within the system itself. In retrospect, it is most surprising that a formal system of entailment rich enough to do biology in is also, at the same time, precisely the system of entailment which biology itself seems internally to manifest.

We reiterate that the system of entailment we have sketched above is very different, and much richer, than that manifested by traditional formalisms, and by the use of such formalisms as images of material nature. As we have noted, the scope of those approaches is restricted entirely to the part corresponding to (1) above. The language of this approach is thus entirely restricted to state and state transiton. These limitations make it impossible even to frame the questions of entailment necessary to encompass the basic questions of biology. To deal with these questions necessitates the introduction of wider processes of entailment; processes which go back to Aristotle, and not to Newton.

It might be noted here that the very idea of Reductionism, on which so much of contemporary biology rests, finds its own expression only within the extended framework of entailment we have presented above. For in this context, Reductionism requires that to understand an operator like f, we must "reduce" it, or disassemble it into parts, *without loss of information*. In the present context, this means we must find a way to *entail f* from these parts. Thus, the parts become the substrate, which we earlier called (?), of the new operator Φ which serves precisely to entail f from these parts. Thus, Reductionism itself rests on something outside the normal limitations of scientific entailment as we now understand it.

5. The Machine

By way of illustration of the above ideas, and also by virtue of its central importance in the traditional approach to biology and other subjects (including, increasingly, physics and mathematics themselves), we shall give a brief discussion of the concept of "the machine."

For a long time, machines had the connotation of being systems which were fabricated, or engineered; artificial or artifactual; the products of design. Thus, to talk as Descartes did, about a *natural* machine, and even more, to identify organisms as falling within this class, involved a provocative extension of the machine concept, which led naturally to the inclusions

organisms ⊂ machines ⊂ mechanisms

to which we alluded earlier. Indeed, nowadays, the term "machine" is used in so many different senses that we must spend a moment to clarify our own usage of this term.

To do this, we shall for a moment leave the natural world, with its pulleys and gears, its engines and switches and computers, and return entirely to formalisms. For it will be the idea of the *mathematical machine* (and more specifically, of the Turing machines) which will allow us to define the concept of machine in general. Specifically, we shall argue that these mathematical or formal machines represent the ultimate *syntactical engines*; the ultimate symbol processors or symbol manipulators. As such, they are intimately connected with the ideas of *formalization* which we discussed above; with the idea that all "truth" can be expressed as syntactic truth. We shall then argue that the activities of these machines can be summed up in a single word: *simulation*. What mathematical machines do, then, is to simulate other machines, including themselves (to the limited extent that this is possible).

We shall then make the connection between this formal world of mathematical machines, and the natural world, by arguing that a *natural system is a mechanism if every model of it can be simulated by a mathematical machine*. By "model", of course, we mean a formal system embodying the properties displayed in Fig. 1 above. This usage will be seen to cover not only the mechanical artifacts we conventionally call "machines", but extend far beyond this, into both animate and inanimate nature. Indeed, the Newtonian paradigm itself, and all those which rest on its epistemological suppositions, make the profound claim that *every natural system is a mechanism* in this sense. The very same assertion, though coming from the formal side, is a form of *Church's Thesis* (cf. Rosen 1962, 1985a).

Let us then proceed with a discussion of the mathematical machines, and more particularly, of the Turing machines. To manipulate symbols, we obviously need symbols. Let us then contrive to fabricate a *finite* set

$$A = \{a_1, a_2, \ldots, a_n\}$$

which will constitute our *alphabet* of discrete, unanalyzable symbolic units. We can obviously, by a process of concatenation, string copies of these symbols into arbitrary finite sequences, which we shall call *words*. These constitute a new set $A^{\#}$, which even possesses a rudimentary algebraic structure; it is the free monoid generated by the alphabet A, under the operation of concatenation.

So far, there is essentially no inferential or syntactic structure. To get some, we must add it in explicitly. Let us then suppose we give ourselves a mapping

$$f: A^{\#} \to A^{\#},$$

which means that we can operate on certain input words w, and obtain corresponding output words $f(w)$. This, as we have seen, is the simplest prototypic inferential structure in any formal system.

But given this single inferential rule, we can "fool" our system into evaluating many other functions for us, and thus carrying out many new inferences. For instance, let us pick an arbitrary word $u_g \in A^{\#}$, and define a new function

$$g: A^{\#} \to A^{\#}$$

by writing

$$g(w) = f(wu_g).$$

Intuitively, this word u_g serves to *program* our system, in such a way as to make it *simulate another system*. In a sense, this prefix u_g serves to *describe* the mapping g to the mapping f, in such a way that any input word which f sees thereafter is treated exactly as g would treat it. In this way, we make the original inferential rule f *simulate* another inferential rule g.

It should, of course, be noted that the above notion of programming and simulation is not restricted to the concatenation operation. It can, obviously, be widely generalized. But it appears even here, and suffices for our purposes.

Let us notice several interesting things about the situation we have described. First, in order for the rule f to *simulate itself*, the corresponding program is *empty*; it is the unit element of $A^{\#}$. Thus, f cannot receive any non-trivial description of itself. In poetic terms, it cannot answer (even if it could pose) the question "who am I?" (I am indebted to Otto Rössler for the above observation.)

Second, let us notice again the dualism between the inferential rule f and the propositions on which it operates. This dualism is parallel to the system-environment dualism to which we referred earlier, although it is not co-extensive with it. It is parallel to the extent that there is no inferential structure in $A^{\#}$ itself; no way to entail, e.g., the next letter of an input word from a given letter. The words in $A^{\#}$ thus are analogous to the acausal environment we described earlier. In the language of mathematical machines, this dualism is expressed as between *hardware*, embodied in the inferential machinery (here the mapping f) and the propositions or words on which it operates (the elements of $A^{\#}$), which constitutes *software*.

Thus, the ability to describe the mapping g to the mapping f which simulates it, or in other words to *program* f to simulate g, amounts to a *literal, exact translation of g from hardware to software*. The possibility of such translation is at the heart of the study of mathematical machines, and hence of machines in general. It is responsible for the strengths of simulation, but also for its profound weaknesses as a tool for exploring formalisms in general, and the material world in particular. For ultimately, it is the supposition that *any* hardware, *any* inferential structure, can be effectively translated into software in this fashion which is at the root of the Newtonian paradigm in science, and of Church's Thesis in the theory of machines.

In the theory of Turing machines, all hardware is embodied in the "reading head" of the machine. This manifests all the inferential structure, all the entailment, which is present in the machine. This "reading head" is imaged in standard Newtonian terms; it has a state set, and a state-transition structure governed by specific mappings, which also determine how it moves, and how its activities are translated into symbols (software) again.

Thus, as stated before, the mathematical machines embody the ultimate in syntax; in symbol processing or word processing. As such, these machines also represent the concrete embodiment of the ideas of formalization described previously, in which all truth could be replaced by syntactic truth; by manipulation of meaningless symbols according to purely syntactic rules (here embodied entirely in the new hardware of the "reading head"); in which all the inferential structure present in the original system could be completely translated into software, where no inferential structure remains at all.

In the theory of mathematical (Turing) machines, we begin with such hardware already in place; it is from this hardware that we obtain the function f. We can then ask what other mappings can be programmed; what other functions g this hardware can simulate. Clearly, if f is computable in this sense, any such g will also be computable; it can be translated entirely into software (program), and its evaluation properties embodied entirely in hardware. We can ask for hardware on which the maximal number of functions g can be simulated; this leads to the idea of the *universal* machines. So the question arises: how much mathematics can these machines do? How successful can the process of formalization be within mathematics itself?

The answer, as has been well known for many years (including the original work of Turing himself), is: these machines cannot do very much. Gödel (1931) was the first to show that arithmetic, the science of number, already could not be formalized in this sense; that no matter how we try to syntacticize its inferential structure, we cannot make arithmetic truth coincide with syntactic truth. That is, there must always be a residuum of arithmetic truth which is of a non-syntactic character (i.e., it is semantic in nature). Stated another way: the inferential processes of arithmetic cannot be completely expressed as software, to be processed symbol by symbol in a Turing machine, without losing (semantic) truths.

These considerations mean in effect that "most" numbers are in this sense non-computable; "most" functions $f: A^\# \to A^\#$ are "non-recursive"; "most" mathematical problems are unsolvable. It also follows that "most" of the inferential processes in mathematics are not expressible in the form of software (programs) to be processed by syntactic hardware, and hence that entailment in mathematics must generally involve an irreducible semantic aspect. If we lose sight of these facts, we also lose the core of mathematics as we know it. We cannot, then, replace the mathematician by a machine of this type, without being able to tell the difference. Already here, the "Turing test" fails.

However, these considerations still leave untouched the central contention of Church's Thesis, or of the Newtonian paradigm: namely, that every *material* system, governed by causality, is a machine; that every model of a material system is simulable or formalizable. This means, we recall, that the inferential structure of the model, and that on which this structure acts, can all be translated into software (program and input) to a machine of Turing type, without loss of truth (in this case, without interfering with the modelling relation of Fig. 1 above).

How can we believe in this possibility, given the feebleness of the Turing machines in the mathematical realm? In fact, the mainstay of this belief resides in the peculiar fact that these machines can always solve differential equations (i.e., differential equations always have computable functions as solutions, just as e.g., polynomial equations always have computable numbers as roots). Thus, insofar as differential equations suffice to image the state-transition sequence in material systems, and insofar as this state-transition sequence is the only mode of entailment allowed in the material world, nothing can happen in the external world which is not completely translatable into software, completely syntacticized, and hence simulable by some mathematical machine.

This is in itself an interesting line of argument, which it is worth taking a moment to consider. At first sight, it appears we can defeat it by exhibiting a material system possessing a model which cannot be expressed as differential equations, or more generally, in which causality is not imaged as state-transition sequence. As we have argued elsewhere (cf. Rosen, 1985b), this is precisely the defining characteristic of what we call *complex systems,* and what distinguishes them from simple systems or mechanisms. Complexity in this sense precisely entails the existence of models which cannot be simulated, and hence directly contradicts Church's Thesis; it shows the existence of an "effective" process ("effective" because it is physical or material) yet cannot be implemented on a Turing machine.

Despite this, it can be argued that the model in question is in some sense a "macro-description," and that there must be an underlying particulate micro-description (e.g. in terms of constituent particles) which *is* simulable, at least in principle. Namely, we can presumably reduce everything to constituent particles, which physics asserts are describable in terms of differential equations, and hence simulable. Thus, by creating the appropriate initial conditions for this micro-description, we could reproduce *any* phenomenon, even one which possesses a non-simulable model. The difficulty here is that the argument merely shifts non-simulability or non-effectiveness to another place (namely, to the inability to reconstruct the macro-description from the micro-one). It would be analogous to argue, for instance, that since the truths of number theory are expressible in symbolic form, all we need to do to reproduce all of them is to simply put the constituent symbols into the right places. But it is precisely this that Gödel showed could not be effectively characterized. We shall now return to the main line of the argument and show that the mode of entailment we called Φ in the preceding section, which we introduced in order to entail an operator or an inferential rule, is already not simulable. More precisely, we shall show that a process which entails an operator cannot be described in terms of a program (software) which can then be simulated in a mathematical machine.

The situation we envisage can, as we recall, be expressed in terms of a perfectly good diagram of mappings

$$A \xrightarrow{f} B \xrightarrow{\Phi} H(A, B).$$

The mapping f represents a respectable Newtonian procedure, which entails material structures or processes in B from corresponding ones in A. Thus, the elements of A and B can be thought of as *states* in the traditional Newtonian sense; the inferential rule or operator f is an equation of motion which sends states (in A) onto states (in B) along traditional trajectories.

But now we have the new inferential rule Φ. If Φ were a traditional Newtonian process, then the image of Φ would also have to be states of something; call it X. But by hypothesis, the "states" x in this image must also be identified with operators or inferential rules in $H(A, B)$. As we have emphasized earlier, however, inferential rules or equations of motion do not get state descriptions in general; they are themselves descriptions of *environments* (of A and B); they represent forces imposed on A by an otherwise unspecified environment which is itself left bereft of necessity or entailment.

Thus, the hypothetical state set X has to be identified with a *(Newtonian) model of the environment of A*. The essential part of this model must be an identification between the properties of states $x \in X$ and the generation of the forces f which operate on A. We can already see that this is an unusual situation; the concept of state is designed to describe the operation of entailment *within* a system, not to generate entailment in another system. It is, in fact, precisely at this point that non-simulability enters the picture.

The problem, then, has become one of attempting to relate two different images or models of the same objects. One description is the traditional Newtonian description of the elements $H(A, B)$ in the form of states of something; the other is a description of these same objects as operators or inferential rules operating on the states of something else (the elements of A). This, as we shall now see, is the crucial feature involved in attempting to *simulate* Φ, or equivalently, to write a program which in some sense describes Φ to a syntactic processor.

If we for the moment assume that Church's Thesis holds, then there is in principle no problem in constructing a program for f; this involves describing to a simulator how f converts (descriptions of) elements of A, entered as input data to the simulator, into (descriptions of) elements of B (the outputs of the simulation). Accordingly, if Φ were of this Newtonian type, a simulator would have to be programmed to convert the outputs of the first simulator into (descriptions of) the states of X. But this would only be the first part of the program. For we would then require another program, to convert the elements of x into *programs* for the operators in $H(A, B)$. In other words, in order to simulate Φ, the simulator would have to be programmed to (a) accept as input data the (descriptions of) elements of B, and (b) to produce as outputs the *programs* which were earlier used to simulate the operators in $H(A, B)$ themselves.

What we claim now is that there can be no program or algorithm for carrying out the basic step of converting state descriptions of the elements $x \in X$ into programs of the type required. Or, stated another way, *there is no general way to convert a state description into an operator description*. To do so would require modes of entailment which are simply not present in the general Newtonian picture.

6. Entailment and Realization

The "origin of life" problem requires us to specify, in one way or another, how to fabricate an organism from things which are not organisms. Everyone knows that this problem is hard, but the considerations of the preceding sections provide us with a more direct insight into why it is hard.

The "origin of life" problem is hard in part because we have thus-far found it impossible to characterize the endpoint at which we are driving. This would, of course, make it hard to fabricate anything. We have argued, however, that the difficulties encountered in attempting to characterize the "living state" are not merely technical; they arise precisely because organisms are complex in our sense, and our science is geared only to deal with the simple. In a nutshell, if Descartes had been right, and organisms were automata, we would be able to express them as software; but we cannot, he was not, and organisms are not.

These considerations imply further that fabrication of organisms is not simply the inverse of reductionistic or syntactic analysis. This is, in fact, true in general; information regarding causal chains (or inferential chains) within a system, or even between systems, does not in general tell us anything about the (quite different) causal chains which culminated in that system.

The two approaches which have thusfar animated the attempt to fabricate living systems do not deal with, or even recognize, the fundamental epistemological difficulties posed by the inadequate inferential and causal structures which they involve. One approach is empirical; it attempts to assemble reductionistically generated fragments of organisms, or homologs of them, and hope to "get lucky" through some process of self-assembly. In this view, the "origin of life" becomes essentially stochastic in nature (the "frozen accident") and outside the reach of causality altogether.

The other approach was that popularized by von Neumann, who espoused the view that life could be simulated. He regarded fabrication or construction as identical with computation, and hence in particular he believed that the existence of a universal computer implied the existence of a universal constructor. But because of the vast difference in causal structure between a simulator and that which it simulates, it is clear that computation and construction are very different things, and that von Neumann's approach has little relation to the material world in general, or biology in particular.

Thus, we need to find a quite different way of approaching the fabrication of organisms. This brings us back again to the (M,R)-systems described above. These systems were meant to capture some irreducible organizational feature characteristic of biological cells, independent of the specific material of which the cell was built. It is thus an inverse to traditional reductionistic modes of analysis, which seek to characterize the matter of which cells are built, independent of the specific organization originally manifested.

The (M,R)-system's organization gives us a way out of one of the impasses facing the traditional approaches to "origins" problems. Specifically, we have a definite endpoint at which to aim; we have not built a cell unless what we have built *realizes* an (M,R)-system. That is, we must be able to build a diagram like that shown in Fig. 1 above, which commutes.

However, to show that a given material system realizes an (M,R)-system is a matter of verification, not fabrication. It still does not tell us how to build or fabricate such a realization; how to establish causal chains culminating in such a system. The essential and surprising point of our preceding discussion, however, is that such a fabrication process is itself an (M,R)-system; the process itself is a kind of realization of that which we seek to fabricate thereby. This is the ultimate reason why the "origin of life" problem is hard; as we have seen, it takes us out of every epistemological category we have inherited from the time of Newton.

References

[1] Bohm, D., *Causality and Chance in Modern Physics*, Harper & Row, New York, 1957.

[2] Bohr, N., *Atomic Theory and Description of Nature*, Cambridge University Press, Cambridge, 1937.

[3] d'Espargnat, B., *Conceptual Foundations of Quantum Mechanics*, Benjamin, Reading, Massachusetts, 1976.

[4] Gödel, K., "On Formally Undecidable Propositions of *Principia Mathematica*," *Monatshefte für Mathematik*, 38 (1931), 173–198.

[5] Kleene, S. C., *Metamathematics*, Van Nostrand, New York, 1951.

[6] Margenau, H., *The Nature of Physical Reality*, McGraw-Hill, New York, 1950.

[7]. Rosen, R., "A Relational Theory of Biological Systems," *Bull. Math. Biophysics*, 20 (1958), 245–260.

[8]. Rosen, R., "A Relational Theory of Biological Systems II," *Bull. Math. Biophysics*, 21 (1959), 109–128.

[9] Rosen, R., "Church's Thesis and its Relation to the Concept of Realizability in Biology and Physics," *Bull. Math. Biophysics*, 24 (1962), 375-393.

[10] Rosen, R., *Anticipatory Systems*, Pergamon, London, 1985a.

[11] Rosen, R., *Theoretical Biology and Complexity*, (Rosen, R., ed.,) Academic Press, Orlando, 1985b.

[12]. Wigner, E., *Symmetries and Reflections*, Indiana University Press, Bloomington, Indiana, 1967.

Causality and Finality in Theoretical Biology: A Possible Picture

RENÉ THOM

Abstract

In this article we describe a necessary condition for science to develop a theoretical side. It lies in the ability to expand a set of empirically observed facts into a larger set of "imaginary" events or virtual facts. No science lacking this extension can be said to be "theoretical." Theoretical biology will exist only insofar as biologists are able to construe a set of theoretical developments generated according to a constructive definition, and to specify how reality propagates among this set of virtual processes. (Such is the case of classical Hamiltonian mechanics, where initial data generate reality inside the set of trajectories describing virtual motion.) That the consideration of virtuality is necessary to science was clearly perceived by Aristotle, who systematized it with the distinction between potentiality and actuality. We show how a set of virtualities might be defined to classify the *Baupläne* of animal organisms, by referring them to three basic spatial axes. And we briefly refer to Aristotle's theory of four causes to explain how efficient and final causes may be subsumed under formal causality (using the notion of "morphogenetic field"). A text from Aristotle's *Politica* particularly illustrative of his general method is given in an appendix.

There is a long-standing debate about whether such a thing as "theoretical Biology" does exist, and even the question of whether it *can* exist has been a subject of considerable doubt among practicians. It appears obvious to them that Biology is—on the contrary—an empirical or an experimental science. Such an attitude raises perhaps the original query: What is a science? If we admit *a priori* that Science is just acquisition of knowledge, that is, building an inventory of all observable phenomena in a given disciplinary domain—then, obviously, any science is empirical. And if we want it to be "experimental," then we have to add to the corpus of natural phenomena the *a priori* boundless profusion of man-made phenomena. Present-day Science fully accepts this viewpoint, not only in Biology but also in all other experimental fields such as Physics and Chemistry. So why bother to theorize?

First of all, it must be observed that no science can do without some way of classifying phenomena. And the processes of classification themselves entail the use of some conceptual tools which, obviously, require us to go beyond the simple perception of facts. That some such "implicit," often naïve, classification of objects or processes is necessary in all sciences, can hardly be doubted. It is the function of natural language to associate with any particular class of phenomena a given word. We may content ourselves with this naïve way of going about it, and continue to store any sort of data in our computer memories. But then the problem of how to shorten the description of phenomena—if only to reduce the cost of storage—will come to the fore.

The concept of physical law is typical of this: a single mathematical formula (like Newton's law of gravitation) may subsume a tremendous amount of phenomenological data (Celestial Mechanics, for instance). But there are no biological *laws* which can, in this respect, be compared with Newton's laws. Biological laws—or theories—are generally nothing more than statements of fact (evolution theory, cellular theory). Here the vocable "theory" means only that we are unable, at a given time, to ascertain the generality of some fact, but only presume it to be general. This a far cry from the mathematical framework of Dynamical theory. There we are given a dynamical system S with a phase space M^{2n} of possible kinetic states, and inside that space a flow X denoting temporal evolution of the state of the system. It means we have been able to embed the naturally existing evolutions into a set of virtual ones, all of which can be ascertained, i.e. constructed according to a canonical procedure; the "law" then prescribes the constraint which selects existing processes from among virtual ones. It is my claim that if a science is *more* than just naïve description, this is due to the fact that it has constructed a set of "virtual" (i.e. imaginary) processes among which it is capable of selecting the real, observable ones. Hence the criterion for true scientificity lies, not in the veracity of observation nor in its accuracy, nor in the use of instruments to help increase the set of observable facts, but in the building of a virtuality of phenomena from which the real ones can be selected by a well-defined logical or mathematical procedure.

If we accept this statement, it is not evident—in Biology at least—that modern Science is more "scientific" than older theories now considered obsolete. Modern Biology has made a fundamental use of instrumentation, essentially in microscopic and biochemical exploration. But this new knowledge does not seem to have led to the formation of a more relevant set of virtualities. Consider Darwinism for instance: the main problem here is how to define, for any species e at time t, the set γ of virtual evolutions out of which the principle of "maximal fitness" will extract the real evolution. This set γ of virtual evolutions is a purely mental object which cannot be constructively defined. When molecular Biology discovered the genome to be made of DNA, the set γ was identified with the set of possible "mutations" of the DNA chain. Initially defined as a set of point mutations, this concept was later enlarged to allow more general changes, such as transpositions, multiple copying of a gene or modifications of messenger RNA maturation processes. But through lack of knowledge about the phenotypical effects of such changes, and their stability, this set remained a purely mental construction with no way of grasping its possible relevance.

The importance of virtuality was recognized long ago by Aristotle, who in his *Physica* emphasized the contrast between potentiality and actuality, i.e., between possible and real. Aristotelian logic, founded on his *Physica,* allowed one to state conditions for reality to propagate among virtual events. In this respect, Aristotelian Physics may be said to contain the first "scientific" theory in the history of mankind. Classical Hamiltonian Physics states the conditions under which, given the flow $X = i$ grad H associated to a Hamiltonian $H : M \rightarrow \mathbf{R}$ in a symplectic manifold M, initial data m_0 will propagate along its trajectory (thus defining the propagation of reality within virtuality).

Coming back to Aristotle, it is interesting to look at the passage in Aristotle's *Politica* [IV, 4, 1290, b25] given in the Appendix. Here the philosopher describes in a proper way what could be called a set γ of imaginary animals, a set defined by all possible combinations of known organs for given functions. Qualitatively, I believe that this is the only rational way of building a theoretical Biology. It is more or less the way I followed in my article (Thom, 1986) after the 1984 Abisko meeting, when I introduced the model of the so-called Physiological Blastula as being the "minimal dynamical model" containing all necessary physiological functions of animality. Here we have first to define the main axes organizing development:

1. The path of the digestive tract inside the organism (from mouth to anus) δ;

2. The general "growth axis" (governing the "elongation" of the embryo) γ;

3. The general direction of the organism's motion (usually taken to be the cephalo-caudal axis) K.

Then we have to identify these three axes with two main directions determined by the environment: the horizontal h, and the vertical v, direction of gravity. For some molluscs, such as the snail *Helix* we have

$\delta = 0$ (almost circular form of the digestive tract)

$\gamma = v + ch$ (oblique, spiral axis)

$K = h$

For terrestrial vertebrates (*Tetrapoda*):

$$K = \delta = \gamma = h.$$

The same for arthropods: all have bilateral symmetry spanned by $[\bar{v}, \bar{h}]$. For Man (bipedal)

$$\gamma = \delta = v$$
$$K = h.$$

This shows for Man a very different Bauplan from that of the other Vertebrates. Although the anatomical differences between Man and Primates are very small, the directions of basic gradients are quite different. (Bipedality may be considered as following a process of frontalization of the head, already found in such vertebrates as the cat, extending the vertical plane from the face to the whole body, thus making the spine vertical.)

In a model such as the Physiological Blastula, the set γ of virtual embryologies is defined as the possible implosion of attractors formed by periodic cycles of the PB. Unfortunately this process of "localizing" attractors in base space is not governed by any precise theory, so that the connection between γ and real development is difficult to establish formally. But at least we are given a formal way of attacking the problem. This sequence of implosions corresponds to what E. Geoffroy Saint Hilaire called in French "les matériaux," giving rise to bones and organs. It is associated with the evolutionary "continuity" of the ontogenetic process (possible discontinuities being "catastrophes"). Seen in this way, the main aim of theoretical Biology is not the prediction, nor the finding, of new facts, but rather the definition of a new domain of virtual biological processes, with the aim of elucidating the way "real" processes propagate among virtual ones.

Here we are quite far from the standard materialistic reductionist viewpoint, according to which all processes in living beings have to be explained by the laws of Physics and Chemistry. Although this perspective is obviously very desirable, the fact remains that we are still a long way from reaching it. If we consider a lump of matter subjected to certain stresses, it is in general extremely difficult to predict accurately what will happen. We may be able to do so for solid bodies; but for multiphasic material media, with complicated inner structures, the situation is quite different. There is, generally, no effective way to predict the behavior of such a medium. As the cytoplasm of a cell has a very complex and still poorly known inner structure, we have little hope of achieving a quantitatively precise description of embryonic development. Those people who believe this has to be done may be right, but they might have to wait for centuries to see their hope realized.

We should restrict our task to obtaining a qualitatively valid model involving the minimal number of factors needed to realize all basic physiological activities. Here we use "necessity by hypothesis" as defined by Aristotle. These functions (like predation for an animal) have to operate if the living being (and its species) are to survive. This may be looked on as "finalistic" reasoning. But as reproduction of life is cyclic, any regulatory process between two points a, b of the cycle can also be looked on as teleological. One has to decompose the whole process into a concatenation of local morphogenetic fields, locally defined by analytical models. The use of such models is a case of "formal causality" in Aristotelian terminology. Efficient causality and final causality may be subsumed under the heading of formal causality. It should be observed that in Aristotelian theory final causality is in fact always conditional. An "act," described as a process by an analytical model, may be stopped short if some impediment intervenes during its course. So the act may fail because of interaction with an accidental factor. When so considered, Aristotelian finality is perfectly compatible with our modern view of (local) determinism. And the use of such finalistic arguments is not at all sterile as it involves the interaction of many factors. The task of choosing the "simplest" way of achieving a given goal is by no means a straightforward one.

Another important aspect of biological theorizing is the absorption of efficient causality into formal causality: how to associate with a discrete cybernetical diagram a continuous flow defined by a (locally) analytical model? Here the catastrophe formalism plays a fundamental role. Two sentences like "The cat catches the mouse" and "Heat melts ice" may be given the same topological interpretation: a representative point crosses the well-defined separatrix in a bifurcation diagram (or in a phase diagram). Here we are dealing with the representation of an act, described by a transitive sentence: A acts on B. As observed by Aristotle, any such action requires that the entities A and B have in common the same genus ($\gamma \acute{\epsilon} \nu o s$): if we look at the genus as an abstract space $\mathcal{G}$ of qualities, then the states of A and B prior to interaction are defined (relative to the quality which spans the genus space) by local "sections" $\sigma: A \rightarrow \mathcal{G} \supset \sigma(A)$, $\sigma: B \rightarrow \mathcal{G} \supset \sigma(B)$. To say that A acts on B means that $\sigma(A)$ occupies the dynamically dominating "sink" in space $\mathcal{G}$. In usual circumstances, $\sigma(A)$ will *attract* $\sigma(B)$ exactly as in Aristotelian (and Newtonian) Physics the Earth's center attracts all heavy sublunary bodies. In a sentence such as "The mother is warming up her child" we have the same process of equalizing the body temperatures of mother and child. Of course a complete classification of the dynamics involved in description of the act is extremely delicate, as sometimes the Patient reacts ("anti-Kinei") against the Agent's action (we then replace Dynamics by Game theory). Moreover, defining the genus space $\mathcal{G}$ sometimes requires statistical averaging on microprocesses exactly as temperature is defined by the mean kinetic energy of molecules in movement. One may also consider the situation where the genus space is spanned by a gradient with a spatial interpretation and the substrate of some biological quality are then ordered with respect to this gradient (e.g. the four Aristotelian elements with respect to terrestrial gravity). This may lead, for instance, to "compartmentalization" of clones in biological morphogenesis; phenomena of this kind are already well documented. Of course these considerations do not solve all the problems attached to biological morphogenesis and embryological development. But at least they offer a path on which to advance.

References

[1] Thom, R. "Organs and Tools: A Common Theory of Morphogenesis" in *Complexity, Language, and Life,* J. L. Casti and A. Karlqvist, eds., Springer, Heidelberg, 1986.

Appendix

(Extract from Pierre PELLEGRIN, *La Classification des Animaux Chez Aristote,* Ed. Les Belles-Lettres, Paris 1982)

Il est pourtant un texte qui, bien qu'il soit extérieur au corpus biologique, semble se proposer comme tâche théorique de *Politique* (IV, 4, 1290b25): Aristote entend y déterminer le nombre des constitutions possibles—c'est-à-dire aussi réelles, car pour le Philosophe, en ce domaine, «presque tout a été découvert » (II, 5, 1264a3)—en examinant la manière dont on recense les différentes sortes d'animaux. Pour la commodité du commentaire nous traduirons ce texte en la décomposant en neuf moments:

[1] C'est comme si nous décidions de prendre ($\lambda\alpha\beta\tilde{\epsilon}\tilde{\iota}\nu$) les sortes ($\epsilon\tilde{\iota}\delta\eta$) de l'animal;

[2] nous déterminerions ($\dot{\alpha}\pi o\delta\iota\omega\rho\dot{\iota}\zeta o\mu\epsilon\nu$) en premier lieu ce qu'il est nécessaire à tout animal d'avoir

[3] (ainsi certains organes sensoriels, et ce qui digère et reçoit la nourriture, comme la bouche et le ventre, et en outre les organes par lesquels chacun d'eux se meut),

[4] et si ces parties sont bien les seules, mais qu'elles présentent entre elles des différences ($\delta\iota\alpha\phi o\rho\alpha\dot{\iota}$)

[5] (je veux dire, par exemple, qu'il y a plusieurs sortes ($\gamma\dot{\epsilon}\nu\eta$) de bouche et de ventre, et aussi d'organes sensoriels et aussi d'organes locomoteurs),

[6] le nombre de leurs conjugaisons ($\sigma\upsilon\zeta\epsilon\xi\epsilon\omega\varsigma$) donnera nécessairement ($\dot{\epsilon}\zeta\ \dot{\alpha}\nu\dot{\alpha}\gamma\kappa\eta\varsigma$) une pluralité de sortes d'animaux ($\pi\lambda\epsilon\dot{\iota}\omega\ \zeta\dot{\epsilon}\nu\eta\ \zeta\dot{\omega}\omega\nu$)

[7] (car le même animal n'est pas susceptible ($\tilde{\delta}\iota o\nu$ + infinitif) d'avoir plusieurs variétés ($\delta\iota\alpha\phi o\rho\dot{\alpha}\varsigma$) de bouche, ni non plus d'oreilles),

[8] si bien que lorsqu'on aura pris toutes les combinaisons possibles ($o\dot{\iota}\ \dot{\epsilon}\nu\delta\epsilon\chi\dot{o}\mu\epsilon\nu o\dot{\iota}\ \sigma\upsilon\nu\delta\upsilon\alpha\sigma\mu o\dot{\iota}$), cela donnera les espèces de l'animal ($\dot{\epsilon}\dot{\iota}\delta\eta\ \zeta\dot{\omega}o\upsilon$), et il y aura autant d'espèces ($\dot{\epsilon}\dot{\iota}\delta\eta$) que de conjugaisons des parties nécessaires;

[9] c'est la même chose pour les constitutions dont nous avons traité.

D'Arcy W. Thomson prête à un tel texte, à l'époque d'Aristote, un caractère d'absolue nouveauté: « It has ever since been a commonplace to compare the state, the body politic, with an organism, but it was Aristotle who first employed the metaphor »[1]. Quoi qu'il en soit la comparaison entre organismes biologiques et sociaux avait alors certainement une force que l'usage lui a fait perdre.

---

[1] D'Arcy W. Thomson (1913), p. 25.

Newton, Aristotle, and the
Modeling of Living Systems

JOHN L. CASTI

Abstract

Here it's argued that the modeling paradigm based upon the Newtonian view of interacting particles and forces is inadequate to capture two of the most essential features of systems involving living agents: self-repair and replication. A comparison of the Newtonian world-view with that of Aristotle is made in order to suggest the missing components in the Newtonian picture. Finally, a system-theoretic framework based upon the functional activities of a living cell is presented, extending the Newtonian picture to include the above critical features of living systems. Some of the mathematical properties of the resulting new processes, termed "metabolism-repair" systems, are given along with a discussion of their use as a modeling paradigm in the life, social and behavioral sciences.

1. The Inheritance of Newton

Newton's world is a world of particles and forces. According to the Newtonian picture, all properties of material systems can be expressed by identifying the particles constituting the system, together with a specification of the outside forces acting upon these particles. Translated into mathematical language, if the system of interest consists of N point particles having positions $q_i(t) \in R^3$ and momenta $p_i(t) \in R^3$ at time t, $i = 1, 2, \ldots, N$, then a consequence of the Newtonian assumptions is the claim that *every* quantity Q of physical interest about the system can be expressed as a function of the positions and momenta of the particles, i.e.,

$$Q(t) = \phi(q_i(t), p_i(t), t), \qquad i = 1, 2, \ldots, N.$$

In particular, if we denote the system state as $x(t) \equiv \big(q(t), p(t)\big)$, where $q(t), p(t) \in R^{3N}$, the Newtonian assumption implies that if we take $Q(t)$ to be the acceleration of the system's particles, then we have the mathematical relationship

$$\text{acceleration} \equiv Q(t) = \ddot{x}(t) = f(x(t), \alpha, t), \qquad \alpha \in R^k,$$

for some appropriate function f, and for some vector of parameters α representing the physical constants in the problem (e.g., the force of gravity, the electric charge, mass, etc.) In the jargon of modern system theory, the Newtonian framework can be thought of as an *input/output* system, where the external forces are the input and the resulting positions and momenta of the particles are the outputs. The foregoing set-up is encapsulated in Newton's famous 2nd Law, and is the mathematical transliteration of the Newtonian vision of how things work, in our universe at least.

Newton's hypothesis that everything knowable about a system is a function of the position and momenta of particles is a truly astonishing claim, matched in its unreasonable effectiveness only by Descartes' earlier claim that the points of our everyday, three-dimensional *physical* space can be identified with the points of the *abstract* mathematical space R^3. And in both these cases, their proponents had to invent an entirely new mathematical apparatus with which to express their remarkable claims. Descartes' analytic geometry and Newton's calculus then served as the vehicles by which their underlying visions of reality could be translated into a mathematical framework, which then served to produce the predictions by which their epistemology and world-view would later be enshrined by the scientific community as *the* framework by which to describe and probe the goings-on of Nature.

Successful as the Newtonian paradigm has been, it is not without its flaws as a faithful description of the universe. In particular, there are a whole host of hidden assumptions "hard-wired"into the Newtonian *Weltanschauung,* assumptions whose elucidation and relaxation later led to major extensions of Newton's vision. For example, Newton assumed that the maximum speed of signal transmission from one particle to another is infinite; Einstein assumed it was finite, and out popped the special theory of relativity. Another Newtonian assumption is that the smallest unit of action is zero; Planck assumed it to be greater than zero and quantized, leading to the Schrödinger equation and the magnificent edifice of quantum mechanics. In another direction, Newton dealt with friction-free, conservative systems; Prigogine said let's look at far from equilibrium, dissipative systems and, lo and behold, a Nobel Prize was waiting in the wings. In fact, it has become something of a cottage industry and a prescription for Nobel fame to pick apart Newton's original framework, identify the tacit assumptions, change these assumptions, and then work out the resulting mathematical and physical implications flowing from the modified equations.

It's important to emphasize the point that virtually all of what is today termed "mathematical physics" or, more generally, "dynamical systems," is deeply rooted in the above Newtonian picture and, hence, ultimately derives from a preoccupation with systems involving particles being acted upon by forces. Not surprisingly, this emphasis leads to a mathematical apparatus that is astoundingly good at describing physical situations in which some system components can be given the role of the "particles," and other features of the situation play the "forces." So, for example, in attempting to generate a "physics" for economics, we see models in which individual consumers are regarded as the particles, while various illusory quantities like utility are regarded as the forces. A similar case arises in population geography, where the individuals in a geographic region are regarded as the particles, and various urban centers are thought of as being masses that exert differing kinds of attractive and repulsive forces upon the individuals. There are many other examples of this sort of physics envy in the social, behavioral and life sciences, all steming from the handful of unqualified modeling successes in physics which themselves trace their success to the basic soundness of the Newtonian paradigm. As convincing as some of these exercises have been, all is not well in paradise.

In my opinion, one of the greatest failures of the Newtonian modeling mind-set takes place when we attempt to invoke the classical framework in order to model processes in the life sciences, particularly those aspects of living systems that are not just physics or chemistry in disguise. Let me elaborate on this point. Many theoretical investigations in biology and medicine involve looking at the physicochemical structure of some biological process, e.g., the transport of chemicals across membranes, the conduction of a nerve impulse, the flow of blood through the kidney, and so forth. For processes of this sort, models based upon the standard Newtonian framework inherited from physics work reasonably well. And why not? After all, there's nothing particularly characteristic of living beings in processes of this kind, and one might just as well be studying the transport of chemicals in an oil refinery or the flow of water through a river network, at least as far as the physical principles and the mathematical apparatus is concerned. But when one turns away from such "pseudo-biological" processes, and starts looking at processes which have no real counterpart in non-living systems, prospects for a successful use of Newton's machine look distinctly bleaker.

The kinds of real biological phenomena that I have in mind here are those involving the most characteristic features of living organisms: self-repair and replication. Since from a biological point-of-view, an input/output process represents a system displaying only a metabolic operation of some sort, perhaps it's not surprising that the Newtonian set-up cannot accommodate non-metabolic, but essential features of living organisms in any natural fashion. After all, the framework was created to deal with situations in which the only thing that was going on was metabolism of some sort, and the characterizing features of living processes such as repair and replication played no role whatsoever in the kinds of particle processes that Newton had in mind. Consequently, if we want to use the Newtonian machinery to model these kinds of situations, we're going to have to do one of two things: either extend the Newtonian set-up to incorporate these additional features, or bend and squeeze and cut the problem so that it can be warped into a form that will conveniently fit into the existing framework. My contention is that virtually all work in mathematical biology, at least to the extent that it involves thinking of a living process as a dynamical system, falls into this second, Procrustean category.

Of course many will dispute the above claim, especially those believers in a resolutely reductionistic view of biological phenomena. To such anti-diluvian souls, everything that is important or that ever could be important about living systems is ultimately reducible to physics and chemistry. And since Newton's vision suffices, roughly speaking, to describe whatever it is we want to know about such physico-chemical situations, there's no need for any kind of theoretical framework other than an *ad hoc* extension or two to the *Principia*. The reductionist creed basically amounts to a religious argument, and wouldn't even be worthy of mention here if it were not so widespread in one form or another throughout all of science, especially in the works of physicists and engineers turned to biology. Let it suffice to say that there has never yet been a knock-down argument supporting any sort of reductionist claims in biology, and I'm not holding my breath waiting for one to appear. Furthermore, it seems to me just plain un- (or even anti-) scientific to cling to the position that when we don't know what to do, we should apply what we do know, and thereby legitimize modeling exercises in biology that are based upon an inadequate, manifestly deficient modeling paradigm originally created to deal with a quite different set of questions. So to my mind at least, there's really nothing left but to take the bull by the horns and ask how we can *naturally* extend the Newtonian picture to incorporate the features that make life live.

One of the necessary conditions for a successful academic career is being able to immediately zero-in, vulture-like, upon the flaws in the arguments of your colleagues or, better yet, your enemies. However, it's one thing to throw stones at the shaky claims of others, but it's quite something else again to try to erect a glass-house of your own. So even if you're willing to swallow the heavy dose of medicine I've prescribed above, it's reasonable to ask just what kind of extensions to Newton's vision I have in mind. I'll come to these matters a bit later on, but let me first try to motivate my arguments by re-examining the world view that Newton overthrew, namely, the Aristotelian theory of causal categories.

2. The World According to Aristotle

The basic goal of mathematical modeling is to be able to answer the question 'why?' According to Newton, a physicist, the corresponding 'because' is given in terms of local interactions involving material particles and unexplainable forces. Aristotle, who was a biologist, had a quite different way of saying 'because.'

In Aristotle's view, the 'why' of things can be described in terms of three basic entities: (i) the material *substance* comprising physical objects, (ii) the abstract or geometric *forms* that objects can assume, and (iii) the processes of *change* by which either the substance or the form may be modified. Thus, Aristotle's 'because' results in four disjoint and inequivalent *causal categories* which, taken together, provide a complete answer to 'why' the world is as it is. These causal categories are:

- *Material* cause—things are as they are because of the physical **matter** of which they are composed;

- *Efficient* cause—things are as they are because of the **energy** that went into making them as they are;

- *Formal* cause—things are as they are because of the **plan** according to which they were built;

- *Final* cause—things are as they are because of the **desire** or **will** of someone to have things take their current state.

Note that in the above scheme of things, material cause corresponds to substance, with efficient cause relating to processes for changing the substance. Similarly, formal cause explains the abstract or geometric form of an entity, with final cause describing how one changes the form. This scheme explains why there are four basic causes in the Aristotelian world-view, and not three or five or 3469.

In the epistemology of Aristotle, all things can be explained by invoking a combination of the four basic causes, with each cause illuminating a different fundamental aspect of the system at hand. We can also interpret these inequivalent causal categories by thinking of each category as being concerned with the manipulation of "something" as depicted in Table 1.

CAUSE	PROPERTY MANIPULATED
Material	Physical Matter
Efficient	Energy
Formal	Information
Final	Desire; Will

Table 1. Aristotelian Causal Categories and Manipulations

As an illustration of how Aristotle would look at things, consider the query: "Why is my car the way it is?" An Aristotelian analysis would say that the car is the way it is because of the steel, glass, rubber and plastic from which it's contructed (material cause); because of the labor that went into its manufacture (efficient cause); because of the design specifications according to which it was built (formal cause); because of the desire for someone (me!) to have a car of this sort, resulting in someone else (the manufacturer) setting up a process for satisfying this demand (final cause). Thus, there are many ways of responding to the question of why a car is a car, and each of these ways is equally valid, but inequivalent and incomplete, insofar as offering a comprehensive explanation for this particular slice of reality. But how does such an epistemology match-up to the Newtonian vision of reality considered earlier?

Interestingly enough, both the Newtonian and Aristotelian explanatory schemes talk about the same thing: a material substance and the process by which this substance can change. However, in the Aristotelian picture substance is not enough; one also needs the idea of form and some kind of dynamic by which one form can be transformed into another. This latter idea is totally absent from the Newtonian picture. In partial compensation, the Newtonian set-up offers a mathematical apparatus by which we can describe both the particles (material cause) and the forces (efficient cause) constituting the Newtonian modeling paradigm. The Aristotelian picture provides no mathematical machinery, only a verbal description of the causes. It's instructive to examine this dichotomy in a bit more detail in order to get a feel for what must be done to extend the Newtonian formalism to accommodate the additional Aristotelian causes.

Newton's 2nd Law is usually written as the differential equation

$$\ddot{x}(t) = F(t), \qquad x(0) = x_0,$$

where x is the state of the system of particles, F represents the unexplained external forces, and x_0 is the initial state of the system. For our purposes, it's more convenient to write this relationship in integrated form as

$$x(t) = x_0 + \int_0^t \phi(s)\, ds,$$

where

$$\phi(s) = \int_0^s F(r)\, dr.$$

Now we can ask the question: "why is the system in the state $x(t)$ at time t?" Newton can give only two answers:

1) The system is in the state $x(t)$ at time t because it was in the state x_0 at time $t = 0$ (material cause);

2) The system is in the state $x(t)$ at time t because of the operator $\int_0^t(\cdots)$ which transformed the initial state to the state at time t (efficient cause).

Thus, the Newtonian framework has neither the need nor the room to accommodate the additional Aristotelian categories of formal and final causation. Some would argue, myself included, that this fact more than any other accounts for the banishment of formal and, especially, final cause from polite scientific conversation for the better part of three centuries. There is just no room for them in the conventional Newtonian framework.

In actuality, even the most die-hard Newtonian ultimately came to recognize, albeit implicitly, that the missing causal categories would somehow have to be grafted on to the classical set-up. You'll recall that in our discussion of the Newtonian framework given in the preceding section, we included a vector of parameters $\alpha \in R^k$ which played the role of setting the scene for the particle system at hand. Thus, the elements of α acted as parameters specifying various important constants in the situation such as particle masses, gravitational constants, electric charges and so on. It's through the specification of such parameters, and their incorporation into the mathematical framework, that formal cause enters by the backdoor into the Newtonian scheme of things. But what about final cause? How does Newton deal with the idea of desire or will? Answer: he doesn't!

When reading Aristotle's account of the causal categories, one is struck by the great significance he attaches to the notion of final cause. In fact, for Aristotle it seems that final cause was just a little more equal than any of the other categories, and he reserved his greatest respect and kindest words for what would today be termed (by Newtonians) "teleology." For the kinds of problems that Newton was concerned with, it appears reasonable to omit final causation from consideration since it's difficult to imagine non-living, material particles having any particular kind of will, volition or consciousness. Thus, Newton and his successors had no need to invoke any of the ideas associated with final cause, notions like goals, plans, will, or even self-reference, in their analyses of physical processes.

From this point of view, it seems easy to understand why the mathematical machinery they employed seemed perfectly adequate to the task at hand, even though it contained no natural way to account for final cause, and even dealt with formal cause in a rather *ad hoc* manner. Unfortunately for the biologist, economist and psychologist, Newton's prescriptions were **too** successful in answering questions in physics, chemistry and engineering, leading to a gradual emergence of the position that it is a breach of scientific etiquette, if not downright un-scientific, to allow anything even faintly smacking of final causation to enter into polite discourse when considering the modeling of any physical process. In other words, if you can't use the methods that work in physics, then you're not doing science. And the "methods" of physics ultimately come down to employment of the Newtonian paradigm.

My claim in this essay is that this attitude is just plain wrong, or at least it's wrong to the degree that the modeling paradigm contains no natural way to include the additional Aristotelian categories of formal and final causation. When one moves away from consideration of non-living systems of the sort dealt with in physics, and enters into the arena of living processes as in biology and the social sciences, then the most pressing questions that emerge are those involving formal and final cause, with matters pertaining to material and efficient cause representing, for the most part, side issues. Thus, my contention is that without a major re-thinking of our underlying modeling paradigm, it's just not possible to even begin to touch the problems of real concern by continuing to limp along with the classical Newtonian-based ideas and tools.

As noted earlier, by clinging to the tools of physics, the only thing that we can hope for is to learn about those aspects of living systems that can be studied independently of the fact that the system is alive rather than dead. To learn about life, something new is needed; it's the system-theoretic nature of this "something new" that I want to consider in the balance of this essay. This is not to say that I advocate throwing out the baby with the bathwater and doing away with the basic Newtonian set-up altogether. Rather, I'll argue that the Newtonian view needs to be *extended* so that it naturally includes not only material and efficient cause, but also the other two Aristotelian categories as well. In a moment, I'll enter into the details of what I mean by "naturally" and the specific nature of the kind of extension I have in mind, but first let me pause for a small, but important digression upon the relationship between causal categories and what I think of as "systems thinking" in theoretical science. By this, perhaps I can indicate a bridge between Greek science and the science of today.

3. Systems, Science and Models

For most of my professional life, I've been associated with departments or institutions whose primary activities (nominally, that is, when they weren't engaged in political squabbling which seemed to be their **real** activity) centered on what in general terms is humorously called "systems analysis." During the course of innumerable conversations with the general public, or even other scientific workers, I have been asked, "just what is systems analysis, anyway?" or "So, you're a system scientist. Well, what *exactly* is system science? Does it have anything to do with computers?" When faced with such embarrassing queries, I mostly mumble some incoherent reply and fervently wish the questioner would disappear, much like the now mythical ether. If I had properly considered my Aristotle, I would have been in a somewhat better position to fend-off these awkward inquiries: *system science is the study of those system properties emphasizing formal and final cause.* While I'm far from convinced that such an "answer" would be any more satisfying than my usual cocktail party mumble to some wide-eyed and innocent "educated layman," it seems to me that the causal explanation of systems thinking goes quite a ways toward distinguishing the "systemness" of any particular study, as well as providing a starting point for identification of the essential character of what we mean by the discipline of "system science."

My claim is that the intrinsic "systemness" of a given situation is characterized by the degree to which the questions and issues of concern focus upon formal and final causation at the expense of material and efficient cause. This classification scheme enables us to separate the traditional disciplinary areas along an axis measuring the "system-determined" nature of their primary activities. By looking at the extent to which the various scientific disciplines emphasize manipulation of information (formal cause) and will or desire (final cause), we can form the admittedly subjective classification indicated in Table 2.

LITTLE	SOME	A LOT
Physics	Engineering	Biology
Chemistry	Medicine	Computer Science
Material Science	Environmental Science	Psychology
Astronomy	Linguistics	Economics
Meteorology		Ecology
		Political Science

Table 2. The "Systemness" of Various Disciplines

In general, the standard Newtonian-based modeling concepts and tools tend to work well for those areas on the left side of Table 2, and as we move toward the right, the Newtonian paradigm becomes progressively less competent to deal with the main questions of concern. This is not surprising at all, since the Newtonian tools were specifically created and refined to address the questions involving material and efficient cause which tend to dominate the so-called "hard" sciences. On the other hand, the "soft" sciences on the right concentrate on issues pertaining to formal and final cause. To expect the standard Newtonian paradigm to work well for such matters is about the same as expecting to be able to cut a steak with a spoon. A spoon is just the thing for dealing with a bowl of soup, but it leaves a lot to be desired when it comes to getting down to cases with a T-bone. So it is also in modeling; we need the right tools for the job at hand, and the standard Newtonian set-up needs some radical extensions before it can competently serve for the "system-determined" sciences. At this juncture, I should pause and clarify just exactly what I mean when I relegate a given discipline to one side or another of the above Table.

As noted earlier, Aristotle felt that complete knowledge of any situation involved understanding the role **each** of the four causal categories plays in bringing about whatever state-of-affairs it is that we're interested in. Thus, in general it would be wrong to think that any given situation can be exclusively explained or understood by appeal to only a subset of the different categories; usually each causal "lamp" illuminates a different facet of the system, and we need all of them to form a complete picture. Consequently, when I say that physics, for example, is dominated by questions of material and efficient cause, all this means is that the questions of primary concern usually focus on matters involving the material structure of objects (molecular and atomic structure and/or elementary particles) and the energy relationships by which these particles are transformed (e.g., S-matrix theory, quark models, and so on). Matters pertaining to formal and final cause don't usually enter into the daily concerns of a modern physicist.

Another interesting case is computer science. Some might argue that an important aspect of computer science involves things like development of new types of semiconductors, exotic memory devices and the like. These are clearly issues involving material and efficient cause, so why is computer science listed on the right-side of Table 2? I have placed it there because to my mind the above matters, while central to computer hardware development, are really questions of physics and engineering not computer science, and the real issues distinguishing computer science as a recognizable discipline revolve about formal and final cause. For example, the design of programming languages, operating systems, distributed computing networks, algorithms and the like are all matters that emphasize both formal cause (information) and final cause (user-friendliness), although they clearly have some material and efficient cause aspects as well, since they must eventually be implemented in physical hardware. But these hardware considerations are not what we usually think of when we think about computer science *as computer science,* so I consider computer science to be a "system-determined" science. Similar arguments apply to the other disciplines listed in Table 2. Now let me get back to the main problem—a modeling paradigm for living systems.

One of the most obvious properties shared by all the disciplines on the right-side of Table 2 is that they focus upon systems in which living agents play the central role. This is true almost by definition with biology and ecology, but holds equally well for computer science too, as just discussed. So if we are to have a modeling paradigm appropriate for such systems, it will have to incorporate those features distinguishing living from non-living systems. And just what are these identifying "fingerprints" of life? Well, it seems that there is a fairly uniform consensus that two of the most important such fingerprints are the capacity to engage in *self-repair,* and the ability to reproduce, i.e., *replication.* Thus, I'll take these two properties as the *desiderata* for the kind of extension to the Newtonian framework that I have in mind. But there are other features that such a modeling framework should include besides just some way of accounting for repair and replication. Let me list them:

• *Extension*—the new framework should include the traditional Newtonian set-up as a subset. It would be foolish to ignore the power of classical dynamical systems as a framework for modeling any kind of process in which change is paramount. Thus, whatever paradigm we propose, it should be reducible to the standard Newtonian view when the processes of life, repair and replication don't enter into the situation.

• *Natural*—the way in which the new framework incorporates repair and replication should follow in a "natural" manner from the classical Newtonian situation. By most accounts in biology and chemistry, life arose on Earth out of originally non-living matter, so I think it's important that whatever extension is made to the Newtonian framework to include repair and replication should obey the same principle. What this means mathematically is that the repair and replication components of our framework should be constructible by **natural** mathematical operations from the metabolic (i.e., Newtonian) part of the formalism. Here I use the term "natural" in the precise sense in which it is understood in mathematics, i.e., the formation of new objects from old by natural operations such as cartesian products of sets, natural projection operations, tensor products and so forth. This property of "naturalness" is important because the literature on modeling living systems, especially in biology and economics, is littered with papers in which the repair and replication aspects of the system are grafted onto the metabolic part in a purely *ad hoc* manner. This is just not good enough; in nature these components arose naturally from the metabolism, and they should also arise naturally in the mathematical formalism.

• *Functional*—since we are concerned with formal and final cause, not material and efficient, our new formalism should focus upon the *functional* activities of the system and basically ignore the *structural* aspects. Thus, the kind of modeling framework I have in mind says nothing about the material elements out of which the system is composed, but only speaks about the way in which these elements are interrelated in order to carry-out the functional activities of the system. As a result, the particular physico-chemical structure of the system is irrelevant; what I'm interested in here is a formalism that addresses the logic of life, not its particular physical structure.

Satisfaction of these requirements, together with the need to have a formalism that explicitly takes into account repair and replication, presents an imposing task. Nonetheless, it's a task that is essential, in my opinion, if real progress is to be made in probing the workings of living systems by mathematical means. It's certainly far from obvious that any such formalism exists that satisfies all of the foregoing requirements, so now let me try to convince you that there is indeed such a framework by giving an account of what have been termed the "metabolism-repair" systems.

4. Abstract Metabolism-Repair Systems

As the role model for a system displaying the sort of features that concern us here, let me consider that most quintessential of living objects, the cell. From a functional standpoint, a living cell engages in three distinct kinds of activities: 1) *metabolic* activity by which the cell carries on its primary function of transforming various chemical compounds into others needed for its existence; 2) *repair* activity in which the cell attempts to counteract disturbances in its operating environment; 3) *reproductive* activity in which the cell acts to preserve its functional activities by building copies of itself. It's important here to note that these functional activities are pretty much independent of the particular physical substrate in which they are carried out. Thus, while a biological cell may act as a small chemical plant transforming one sort of chemical into another, an economic system may perform exactly the same functional activities but with no chemicals of any sort. Instead the economic system transforms money or labor or some other sort of raw "material" from one form to another. It's in this sense that I claim that "systemness" consists of emphasizing the formal and final cause over the material and efficient.

An abstract formal mathematical structure suitable for capturing the essence of the three distinct types of cellular activity was proposed some years ago by Rosen under the rubric "metabolism-repair" systems [1–3]. This formalism encompasses all of the requirements noted above but without sacrificing flexibility in formulation, thereby allowing considerable leeway in the details by which the processes of metabolism, repair and replication make their appearance in the overall structure. But rather than speak in such vague generalities, let me briefly sketch the main steps in Rosen's development leaving the technical details for the next section.

We begin with the simple observation that the cell has evolved to perform some kind of metabolic function. What this means is that the business of any given cell is to transform some collection of chemical inputs into a design metabolic output. Thus, if we let Ω represent the set of possible environments that the cell may face, while letting Γ denote the set of possible cellular outputs, then for any given cell there is a design environment, call it ω^*, such that the cellular machinery f^* maps ω^* to the design output $\gamma^* \in \Gamma$. In general, of course, there may be a whole range of acceptable environments ω^*, and correspondingly acceptable outputs γ^*. But for sake of simplicity and ease of exposition, we shall assume that each cell has only a single design environment $\omega^* \in \Omega$, which is transformed by the cellular machinery into a single design output $\gamma^* \in \Gamma$. So when all is working according to plan, the cell accepts the design environment ω^* and processes it according to the basal metabolism f^*, producing the design output γ^*. For future reference, note here that the metabolic map f^* belongs to the set $H(\Omega, \Gamma)$ of possible cellular metabolic mechanisms. This set is determined by various constraints of a physical and chemical nature on just what the cell can do, and may differ widely from cell type to cell type.

It's possible to compactly summarize the cellular basal metabolism by saying that it consists of a map

$$f^* : \Omega \to \Gamma,$$
$$\omega^* \mapsto \gamma^*$$

In passing, let me point out that this is exactly the set-up from Newtonian dynamics when those dynamics are expressed in input/output form. That is, the usual Newtonian picture as expressed earlier by the equation

$$\ddot{x} = F(t), \qquad x(0) = x_0,$$

is precisely a system of the above metabolic type when we take $\omega^* \equiv F(t)$, and let the system output $\gamma^* \equiv x(t)$. In this setting, the input/output map f^* corresponds to the integral operator obtained earlier by expressing the above dynamical process in integrated form. Consequently, we've now reached the point at which the standard Newtonian framework quits. How can we make the processes of repair and replication emerge out of the metabolic "data" Ω, Γ, and $H(\Omega, \Gamma)$?

Functionally speaking, the role of the repair operation is to stabilize the cellular metabolic activity in the face of disturbances to the ambient environment ω^* and/or to the internal metabolic processing operation f^*. For our purposes, we'll suppose that the way this is accomplished is that the cell "siphons off" some of its metabolic output, and then processes this part of the output by a repair mechanism in order to produce a new metabolic map f, which is then used to process the next environmental input ω. Thus, we postulate the existence of a repair map

$$P_{f^*} : \Gamma \rightarrow H(\Omega, \Gamma)$$
$$\gamma \mapsto f$$

Here we include the subscript f^* to explicitly indicate that the role of the repair map is to stabilize the cell's metabolic operation. Consequently, we have the obvious boundary condition on P_{f^*}: "if it ain't broken, don't fix it," which we can translate into the mathematical requirement

$$P_{f^*}(\gamma^*) = f^*, \tag{$*$}$$

indicating that when the cell is operating as it should (i.e., producing the design output γ^*), the repair mechanism should provide the basal metabolism f^* to process the next input. We defer to the next section consideration of how to construct the repair map P_{f^*} in a mathematically "natural" manner from the given design metabolism. For now, let's turn to the third cellular function, replication.

Biologically, the replication operation is a complicated processing of the cellular output in a special way in order to create a copy of the genetic information contained in the cellular DNA. In our abstract setting, the internal model that the cell has of its own structure is embodied in the condition ($*$), which serves as a self-referential model of what the cell is supposed to be doing. Somehow we have to find a replication operator which accepts some subset of the cellular inputs, outputs, metabolism and repair operations, and produces a new active genetic component, which for us is represented by the repair map P_{f^*}. There are probably several different ways to do this, but it turns out to be mathematically very convenient to assume that the replication component accepts some part of the metabolic machinery f, and then processes this into the new repair map P_{f^*}. Thus, we postulate a replication map

$$\beta_{f^*} : H(\Omega, \Gamma) \rightarrow H(\Gamma, H(\Omega, \Gamma))$$
$$f \mapsto P_f$$

Again we have the obvious boundary condition

$$\beta_{f^*}(f^*) = P_{f^*}. \tag{$\dagger$}$$

Let's summarize the development thus far by means of the following diagram:

$$\Omega \overset{f}{\underset{\text{metabolism}}{\rightsquigarrow}} \Gamma \overset{P_f}{\underset{\text{repair}}{\rightsquigarrow}} H(\Omega, \Gamma) \overset{\beta_f}{\underset{\text{replication}}{\rightsquigarrow}} H(\Gamma, H(\Omega, \Gamma))$$

This diagram of sets and maps, together with the boundary conditions (∗) and (†), forms an abstract representation of a (M, R) (metabolism-repair)-system. As noted earlier, the Newtonian part of this diagram appears if we neglect the repair and replication operations, concentrating solely upon metabolism. It is for this reason that I refer to the usual Newtonian systems of physics and engineering as "metabolism-only" processes. To see whether this abstract framework meets the requirements stated earlier as a candidate paradigm for modeling living systems, it will be necessary to put some mathematical meat onto the abstract skeleton of sets and relations by endowing these sets and maps with specific structure. But before entering into these slightly deeper technical waters, it's useful for gaining insight into the unusual (from a Newtonian perspective) properties of (M, R)-systems to examine the operation of the repair component in this more abstract setting.

Suppose the cell's design environment is ω^*, with a corresponding basal metabolism f^* that produces the output γ^* when all is working as it should. Under these design conditions, the cellular repair machinery is the map P_{f^*}. Now imagine that there is some sort of disturbance to the system having the effect of changing the metabolic machinery from f^* to f. In such a situation, the next input ω^* will be processed as

$$f(\omega^*) = \gamma \neq \gamma^*,$$

which in turn will be processed by the repair machinery originally set-up to produce f^*. This processing will yield

$$P_{f^*}(\gamma) = \hat{f}.$$

There are now three possibilities to consider:

1) $\hat{f} = f^*$: in this case the system's original basal metabolism has been *repaired* (restored) by the repair component, and everything returns to the state it was in before the metabolic disturbance;

2) $\hat{f} = f$: here we have the situation where the repair component *stabilizes* the cell at the new metabolic level f;

3) $\hat{f} = \bar{f} \neq f^*$ or f: this result means that the repair component produces a metabolic map that neither restores the original basal metabolism, nor stabilizes the cell. In effect, this outcome means that the cell "hunts" in the set $H(\Omega, \Gamma)$ of possible metabolisms by going through a sequence of metabolisms $\{f^{(i)}\}, i = 1, 2, \ldots$. If there exists some $N < \infty$ such that

$$f^{(N)} = f^{(N-k)}, \qquad k < N,$$

then the cell stabilizes at the new metabolic cycle $f^{(k)}, f^{(k+1)}, \ldots, f^{(N)}$; if there is no such N, the cell's metabolism displays aperiodic behavior.

Remark: The above types of behavior correspond exactly to the three different types of limiting behavior for a dynamical system—an equilibrium point, a limit cycle or aperiodic trajectories. But note that the last case can occur only if the set $H(\Omega, \Gamma)$ is infinite.

Obviously, it will be of considerable interest to know when one of the first two cases occurs, since these are the only situations in which the cell's functional activity can be maintained. Now let's consider the complementary question when the basal metabolism f^* is left unchanged, but there is a disturbance to the design environment so that $\omega^* \longrightarrow \omega$.

In the case of a disturbance to the environmental inputs, the condition for stable operation of the cell is that either

$$f^*(\omega) = f^*(\omega^*)$$

or

$$P_{f^*}(f^*(\omega)) = \bar{f},$$

with

$$\bar{f}(\omega) = \gamma^*.$$

In the first case, the environmental disturbance is invisible to the cellular basal metabolism; in the second case, the cell's repair machinery is able to compensate for the environmental change by modifying the basal metabolism in just the right way to maintain the design output. But note that the cell's entire metabolic activity would be permanently altered to $\bar{f}$ only if

$$P_{f^*}(\bar{f}(\omega)) = \bar{f}.$$

On the other hand, if we had

$$P_{f^*}(\bar{f}(\omega)) = f^*,$$

then the cell's metabolism would only undergo periodic changes cycling back-and-forth between f^* and $\bar{f}$. It's obviously important to understand just what sorts of environmental disturbances can be neutralized by cellular repair machinery, and what kinds of disturbances cause the cell to undergo periodic or aperiodic (unstable) behavior. We'll consider these matters in detail in the next section.

While we have only indicated here some of the types of questions
that arise in connection with repair, there are a host of related questions that emerge when we consider replication. For example, can a
"mutation" $f^* \longrightarrow f$ generate a permanent change in the repair (genetic) map P_{f^*} by means of influencing the replication map β_{f^*}? This
is an issue relating to the circumstances under which Lamarckian-type
inheritance can occur. Even this simple matter already shows a major departure from the standard Newtonian-based stability questions
forming the bread-and-butter core of classical dynamical systems investigations. Here we're not so much interested in what an uncontrolled
system will do, but rather our focus of concern is upon what a system
can do to maintain its functional integrity in the face of ambient disturbances of various sorts. These are exactly the kinds of issues that
are important for living systems engaging in a constant battle with
the outside world to preserve their viability. So with these ideas in
mind, let's turn away from abstractions and philosophy, and look in
the direction of specifics to examine the (M, R)-systems in a definite
mathematical setting—when everything in sight is linear.

5. A Crash Course in Linear System Theory

In order to address the kinds of questions raised above at anything
more than a superficial level, it's necessary to impose more specific
mathematical structure on the initial cellular data: the set of possible
cellular environments Ω, the set of potential cellular outputs Γ, and the
set of feasible metabolisms $H(\Omega, \Gamma)$. To see everything in the simplest
setting possible, here we will assume that all these objects are *linear*.
Thus, we take

$$\Omega = \{\text{finite sequences of vectors in } R^m\},$$
$$\Gamma = \{\text{sequences of vectors in } R^p\},$$
$$H(\Omega, \Gamma) = \{\text{sequences of } p \times m \text{ real matrices}\}.$$

With this set-up, each element $\omega \in \Omega$ has the form

$$\omega = (u_0, u_1, \ldots, u_N), \qquad u_i \in R^m, \qquad N < \infty,$$

while each output $\gamma \in \Gamma$ can be expressed as

$$\gamma = (y_1, y_2, y_3, \ldots), \qquad y_i \in R^p.$$

Note here that for technical reasons, there are only a finite number
of non-zero inputs but that the output sequence is potentially infinite.

With these vector space structures for Ω and Γ, a cellular metabolism f can be identified with the sequence of matrices

$$f \longleftrightarrow \{A_1, A_2, A_3, \ldots\}, \qquad A_i \in R^{p \times m}.$$

In view of the linearity assumption on the map f, given a particular input/output pair (ω, γ), we have the relation

$$y_t = \sum_{i=0}^{t-1} A_{t-i} u_i, \qquad t = 1, 2, \ldots . \tag{I/O}$$

Note that to respect causality, the above framework assumes that the first component of the cellular output appears one time-unit after application of the first component of the input.

The foregoing relation (I/O) relates the cellular output to input; in system-theoretic terms this is called an *external* description of the cell. Usually in system theory, we're also interested in what kind of *internal* mechanism is at work in the cell's interior acting to process the observed input into the measured output. For this we need what's called an internal model, one whose input/output behavior agrees with that actually seen. Mathematically, such a model takes the form

$$\begin{aligned} x_{t+1} &= Fx_t + Gu_t, \qquad x_0 = 0, \qquad x_t \in R^n, \\ y_t &= Hx_t, \qquad t = 0, 1, 2, \ldots, \end{aligned} \tag{Σ}$$

where F, G and H are matrices of sizes $n \times n$, $n \times m$ and $p \times n$, respectively. The integer n defines the dimension of the system's state-space, and is determined by the properties of the original input/output description. In the above set-up, the input and output sequences are exactly the same as that in the given (I/O) description. Thus, determination of an internal model involves finding a triple of matrices $\Sigma = (F, G, H)$ and an integer n such that the relation (Σ) holds. It's clear that the internal model (Σ) contains far more information about the system's workings than does the external model (I/O).

The trickiest part of this external $\longrightarrow$ internal passage is the determination of the number n, representing the size of the internal state-space needed to accommodate the given (I/O) pattern. In general, we want an internal model that is as small as possible, consistent with the requirement that its behavior agree with what has been observed. It can be shown that from a technical point of view, this invocation of Occam's Razor translates into the conditions that the the model (Σ) be

what is termed "completely reachable" and "completely observable."
The first condition means simply that by applying a suitable input, it is
possible to transfer the system state from the origin to any point of R^n
in at most n time-steps. The second condition involves the amount of
information contained in the observed output about the system's initial
state. If Σ starts in an arbitrary initial state $x_0 \neq 0$, then the system
is completely observable if it is possible to uniquely identify x_0 from an
output sequence $\{y_1, y_2, \ldots\}$ containing at most n observations. If Σ is
both completely reachable and completely observable, then we call the
system "canonical." Mathematically, these conditions can be written
in several ways, perhaps the most straightforward being:

$$\Sigma \text{ is } \textit{completely reachable} \text{ if and only if}$$
$$\text{rank } [G|FG|F^2G|\cdots|F^{n-1}G] = n;$$

$$\Sigma \text{ is } \textit{completely observable} \text{ if and only if}$$
$$\text{rank } [H|HF|HF^2|\cdots|HF^{n-1}]' = n.$$

In summary, the construction of a canonical model Σ from the
input/output description (I/O) involves finding a triple of matrices
$\Sigma = (F, G, H)$ such that:

1) The input/output behavior of Σ agrees with that of (I/O), i.e.,
$A_t = HF^{t-1}G$ for all $t = 1, 2, \ldots$;

2) Σ is completely reachable and completely observable, i.e., canonical.

These conditions constitute what in system theory is usually termed
the Realization Problem. (*Remark:* The most general form of the
Realization Problem assumes that it is also necessary to determine the
initial state x_0 of the internal model. For the sake of exposition, we
have assumed that $x_0 = 0$. This assumption is easy to justify in the
linear case considered here. For nonlinear dynamics, the situation is
somewhat more delicate.)

The preceding discussion omits consideration of a crucial point: If
we're given the (I/O) data $f = \{A_1, A_2, A_3, \ldots\}$, how do we determine the dimension n and the matrices $\Sigma = (F, G, H)$ of a canonical
realization? It turns out that there are now many algorithms for doing
this, all of them in one way or another a version of the classical invariant factor theorem of linear algebra. Since it's important for what
follows, let me quickly sketch one of these methods, the Ho Realization
Algorithm.

Arrange the (I/O) data $f = \{A_1, A_2, A_3, \dots\}$ in the following Hankel array:

$$\mathcal{H} = \begin{pmatrix} A_1 & A_2 & A_3 & \cdots \\ A_2 & A_3 & A_4 & \cdots \\ A_3 & A_4 & A_5 & \cdots \\ \vdots & \vdots & & \ddots \end{pmatrix}$$

It can be shown that Σ is a canonical model for the data f, if and only if $\dim \Sigma = \operatorname{rank} \mathcal{H}$. Thus, the data admits a finite-dimensional canonical model if and only if $\mathcal{H}$ has finite rank. From now on, **assume** this to be the case with $\dim \Sigma = n < \infty$ (we'll consider later what to do if this assumption can't be justified). Now the problem is to find matrices F, G, H satisfying the relationship

$$A_t = H F^{t-1} G,$$

with the pair (F, G) being completely reachable, and the pair (F, H) being completely observable.

Let $\mathcal{H}_{rr}$ be an $r \times r$ principal submatrix of $\mathcal{H}$ such that rank $\mathcal{H}_{rr} = n$. Define the operation $\sigma \mathcal{H}_{rr} = $ the "left-shift" of $\mathcal{H}_{rr}$, i.e., shift each entry of $\mathcal{H}_{rr}$ one position to the left. Further, determine matrices P and Q such that

$$P\mathcal{H}_{rr}Q = \begin{bmatrix} I_n & 0 \\ 0 & 0 \end{bmatrix}, \qquad I_n = n \times n \text{ identity.}$$

Such matrices P and Q always exist by virtue of the assumption on the rank of $\mathcal{H}$. We can now write down the matrices of a canonical realization Σ as

$$F = \mathcal{R}_n P \sigma \mathcal{H}_{rr} Q \mathcal{C}_n,$$
$$G = \mathcal{R}_n P \mathcal{H}_{rr} \mathcal{C}_m,$$
$$H = \mathcal{R}_p \mathcal{H}_{rr} Q \mathcal{C}_n,$$

where the "editing" matrices $\mathcal{R}_i$ and $\mathcal{C}_j$ have the actions: "keep the first i rows" and "keep the first j columns," respectively.

Example: The Natural Numbers—suppose we have a cell whose basal metabolism is to accept the design input

$$\omega^* = (1, 1, 0, 0, \dots),$$

and produce a design output consisting of the natural numbers, i.e.,

$$\gamma^* = (1, 2, 3, 4, \ldots).$$

It's a simple exercise to verify that in this case, the basal metabolism f^* is given by

$$f^* = \{A_1, A_2, A_3, \ldots\} = \{1, 1, 2, 2, 3, 3, 4, 4, \ldots\}.$$

Thus, in this situation we have a cell with a single-input channel $(m = 1)$ and a single-output channel $(p = 1)$, with each $A_i \in R^{1 \times 1}$. Furthermore, because of the regular structure in f^*, it's an easy matter to verify that for the Hankel array $\mathcal{H}$ we have rank $\mathcal{H} = 3 = n^*$. Going through the calculations outlined above for the Ho Algorithm, we are soon led to the canonical system

$$F^* = \begin{pmatrix} 0 & 1 & 0 \\ 1 & -1 & 1 \\ 1 & -2 & 2 \end{pmatrix} \qquad G^* = \begin{pmatrix} 1 \\ 1 \\ 2 \end{pmatrix}, \qquad H^* = (1 \quad 0 \quad 0).$$

Thus, the canonical dynamical system realizing this cell is then

$$x_{t+1} = \begin{pmatrix} 0 & 1 & 0 \\ 1 & -1 & 1 \\ 1 & -2 & 2 \end{pmatrix} x_t + \begin{pmatrix} 1 \\ 1 \\ 2 \end{pmatrix} u_t, \qquad x_0 = 0,$$

$$y_t = (1 \quad 0 \quad 0) x_t, \qquad t = 0, 1, 2, \ldots \qquad (\Sigma)$$

This concludes the very brief summary of ideas and tools from linear system theory needed for studying (M,R)-systems. Those readers wanting further details on all the items discussed here, along with a treatment of many other aspects of linear systems, are invited to consult the references [4–6] for a full account. Now let's get back to the (M,R)-systems.

6. Linear (M,R)-Systems: Repair

With the above machinery of linear systems at hand, we can finally address the new components of the (M,R)-paradigm: repair and replication. Suppose we have a given cellular basal metabolism f^*, and we want to determine the corresponding repair map P_{f^*}. By virtue of the condition that P_{f^*} should reproduce f^* when the design input ω^* and the basal metabolism are unchanged, we have that $P_{f^*}(\gamma^*) = f^*$. Also by the linearity assumptions, we must have

$$P_{f^*} \longleftrightarrow \{\mathcal{R}_1^*, \mathcal{R}_2^*, \mathcal{R}_3^*, \ldots\},$$

for some linear operators $\{\mathcal{R}_i^*\}, i = 1, 2, \ldots$. Putting these observations together, the (I/O) relationship for the repair system is

$$w_\tau^* = \sum_{i=0}^{\tau-1} \mathcal{R}_{\tau-i}^* v_i^*, \qquad \tau = 1, 2, \ldots,$$

where the pair (w_i^*, v_i^*) are the output and input to the repair system, with the elements $\mathcal{R}_j^*$ being linear maps determined by the basal metabolism. The condition given earlier on P_{f^*} implies that we must have

$$w_\tau^* = A_\tau^* \qquad \text{and} \qquad v_\tau^* = y_{\tau+1}^*.$$

Note here that we have introduced the new time-scale τ to indicate the fact that the repair system may, and often does, operate on a quite different temporal level than the metabolic part of the system. In light of the linearity assumptions as well as the dimensions of the elements w_τ^* and v_τ^*, the elements $\{\mathcal{R}_i^*\}$ defining the repair map P_{f^*} must have the form

$$\mathcal{R}_j^* = [B_{j1}^* | B_{j2}^* | \cdots | B_{jp}^*], \qquad B_{js}^* \in R^{p \times m}, \qquad j = 1, 2, \ldots.$$

Here recall that p and m are the number of outputs and inputs to the cellular metabolism, respectively.

Remarks:

(1) If we write each A_i as

$$A_i = [A_i^{(1)} | A_i^{(2)} | \cdots | A_i^{(m)}], \qquad A_i^{(j)} \in R^p,$$

then it's natural to think of the "complexity" of each component of the metabolic map as being proportional to $O(pm)$, the number of quantities that have to be specified in order to determine A_i. Similarly, the complexity of each element $\mathcal{R}_j$ of the repair map is $O(p^2 m)$. Thus, by whatever actual measure of complexity one chooses, the repair map is more complex than the metabolic map, an observation that has often been made on biological grounds but which we see here emerging through natural mathematical requirements.

(2) A simple calculation shows that the assumption $\dim \Sigma = n < \infty$ for the cellular metabolic system implies that the set $\{A_1, A_2, \ldots, A_{2n}\}$ is linearly dependent (this follows from elementary properties of the Hankel array $\mathcal{H}$). A similar computation shows that the condition $\dim \Sigma < \infty$ also implies that the canonical model for the (I/O) behavior of the repair system will have a dimension $n_P \leq n < \infty$. Thus, we can again apply Ho's Algorithm to produce a canonical model from the repair sequence $\{\mathcal{R}_1, \mathcal{R}_2, \ldots\}$, obtaining a system $\Sigma_P = (F_P, G_P, H_P)$ characterizing the internal dynamics of the repair system.

Example: The Natural Numbers (cont'd.)

Returning to the example started in the last section, since we must have $P_{f^*}(\gamma^*) = f^*$, after a bit of algebra we find that the repair (I/O) elements are

$$\mathcal{R}_i^* = \begin{cases} [\ 1], & i \text{ odd}, \\ [-1], & i \text{ even}. \end{cases}$$

This leads to the associated Hankel array

$$\mathcal{H}_P^* = \begin{pmatrix} 1 & -1 & 1 & -1 & 1 & \cdots \\ -1 & 1 & -1 & 1 & -1 & \cdots \\ 1 & -1 & 1 & -1 & & \cdots \\ \vdots & \vdots & \vdots & \vdots & \vdots & \cdots \end{pmatrix}$$

It's clear from the above pattern that rank $\mathcal{H}_P^* = n_P^* = 1$, leading to the canonical realization of the repair system as $\Sigma_P^* = (F_P^*, G_P^*, H_P^*)$ with

$$F_P^* = [-1], \quad G_P^* = [1], \quad H_P^* = [1].$$

The repair dynamics then become

$$z_{\tau+1}^* = [-1]\, z_\tau^* + [1]\, v_\tau^*, \qquad z_0^* = 0,$$
$$w_\tau^* = [1]\, z_\tau^*, \qquad \tau = 0, 1, 2, \ldots .$$

From our earlier remarks, we connect this system with the metabolic system via inputs and outputs as $w_\tau^* = A_\tau^*$, $v_\tau^* = y_{\tau+1}^*$.

With the above preliminaries in hand, let's turn back to a consideration of the main function of the repair map: to stabilize the cellular input/output behavior in the face of ambient fluctuations in either the environment, or in the metabolic machinery itself. As seen in an earlier section, there are two cases to consider here: changes in ω^* and changes to the basal metabolism f^*. Consider the second case first.

Case I: A fixed environment ω^* and a fixed genetic machinery P_{f^*}, with a variable metabolism f.

In this case, we are concerned with departures in the metabolic machinery from the basal metabolism f^*, and we want to characterize all those new metabolisms f for which the repair system will stabilize the cell in the sense of either restoring the basal metabolism f^*, or fixing the metabolism at the new level f.

To study this question, it's useful to introduce the map

$$\Psi_{\omega^*, f^*} : H(\Omega, \Gamma) \to H(\Omega, \Gamma),$$
$$f \mapsto P_{f^*}(f(\omega^*))$$

Those metabolisms f which are fixed by the repair map correspond to the fixed points of Ψ, while those metabolic disturbances f that are "repaired" correspond to the problem of finding those f for which

$$\Psi_{\omega^*, f^*}(f) = f^*.$$

Note that by construction we must have

$$\Psi_{\omega^*, f^*}(f^*) = f^*,$$

so that f^* is a trivial fixed point of Ψ, as is the null metabolism $f = 0$ by virtue of the fact that Ψ is a linear map, being the composition of the two linear maps f and P_{f^*}.

The operator Ψ_{ω^*, f^*} can be represented by the infinite matrix

$$\Psi_{\omega^*, f^*} = \begin{bmatrix} \Psi_{11}^* & \Psi_{12}^* & \Psi_{13}^* & \cdots \\ \Psi_{21}^* & \Psi_{22}^* & \cdots & \\ \vdots & \vdots & \vdots & \vdots \end{bmatrix},$$

where each $\Psi_{ij}^* \in R^{p \times p}$, $i, j = 1, 2, \ldots$. The conditions for stabilization of the cellular metabolism can now be expressed in terms of the above matrix Ψ_{ω^*, f^*} as

METABOLIC REPAIR THEOREM. *(1) The metabolic perturbation $f = \{A_1, A_2, \ldots\}$ will be stabilized by the repair machinery if and only if the vector $(A_1, A_2, \ldots)'$ is a characteristic vector of Ψ_{ω^*, f^*} with associated characteristic value 1.*

(2) The metabolic perturbation f will be repaired if and only if f has the form
$$f = f^* + \ker \Psi_{\omega^*, f^*}.$$

In other words, f will be repaired if and only if

$$(A_1 - A_1^*, A_2 - A_2^*, \ldots) \in \ker \Psi_{\omega^*, f^*}.$$

So we see that the answer to our basic question of just what sorts of metabolic disturbances can be stabilized comes down to the linear-algebraic problem of determining the kernel of the map Ψ_{ω^*, f^*}. I'll consider this matter in a moment, but first let's reflect upon just what the Metabolic Repair Theorem *really* says about the possibility of stabilizing the cellular metabolism.

Given a random perturbation f, the chance that the original basal metabolism will be restored is directly proportional to the size of the kernel of the map Ψ_{ω^*, f^*}, as we can see by the second part of the Metabolic Repair Theorem. Thus, for restoring the original basal metabolism, we would like to have the kernel of Ψ_{ω^*, f^*} "large;" on the other hand, the first part of the Theorem says that if we want to have a reasonable likelihood that the cellular metabolism will "lock-on" to the new metabolism f, then we had better have the kernel of Ψ_{ω^*, f^*} "small." These two situations are, of course, mutually incompatible. Thus, there is a trade-off between having a high likelihood of being able to restore the original metabolism f^* (kernel Ψ_{ω^*, f^*} large), and a high likelihood of being able to stabilize the cellular metabolism at the new level f (kernel Ψ_{ω^*, f^*} small). It seems reasonable to suppose that Nature has arranged some sort of compromise between these two extremes, probably dictated by the degree of difficulty in arranging the physical implementation of the two cases, as well as the likelihood of typical perturbations f being "close" or "far" from the design metabolism f^*. Since the cellular output is a continuous function of the metabolism, my own guess is that if the typical kinds of disturbances f are close to f^*, then the cell will opt for a strategy that makes the kernel of Ψ_{ω^*, f^*} small, thereby being able to easily lock-on to the perturbed metabolism. On the other hand, if the perturbations can routinely be major departures from f^*, then it seems to me that the alternate strategy would be an advantageous survival trait, hence favored on evolutionary grounds. Now let's explore how to actually calculate the critical quantity $\ker \Psi_{\omega^*, f^*}$ from the original basal metabolism f^*.

From the component representation of the repair map P_{f^*}, we have

$$A_\tau = \sum_{j=0}^{i} \sum_{i=0}^{\tau-1} [\mathcal{R}_{\tau-i}^{*(1)} | \cdots | \mathcal{R}_{\tau-i}^{*(p)}] A_{i-j+1} u_j^*, \qquad \tau = 1, 2, \ldots . \qquad (*)$$

This is clearly a triangular (in fact, Toeplitz) representation, since A_τ depends only upon the elements $A_1, A_2, \ldots, A_\tau$. On the other hand, we also have the fixed-point requirement

$$
\begin{bmatrix}
\Psi_{11}^* & \Psi_{12}^* & \Psi_{13}^* & \cdots \\
\Psi_{21}^* & \Psi_{22}^* & \Psi_{23}^* & \cdots \\
\vdots & \vdots & \vdots & \cdots
\end{bmatrix}
\begin{bmatrix}
A_1^* \\
A_2^* \\
\vdots
\end{bmatrix}
=
\begin{bmatrix}
A_1^* \\
A_2^* \\
\vdots
\end{bmatrix},
\qquad (**)
$$

for some triangular choice of the $\{\Psi_{ij}^*\}$. But the relationship $(*)$ must hold for **all** sequences $\{A_i\}$, in particular, when $A_i = A_i^*$. Thus, imposing this condition and equating the relationship $(*)$ to $(**)$, we obtain the condition

$$
\Psi_{\tau 1}^* A_1^* + \Psi_{\tau 2}^* A_2^* + \cdots = \sum_{j=0}^{i} \sum_{i=0}^{\tau-1} [\mathcal{R}_{\tau-1}^{*(1)} | \cdots | \mathcal{R}_{\tau-i}^{*(p)}] A_{i-j+1}^* u_j^*,
$$

for $\tau = 1, 2, \ldots$. A slightly technical argument given in [7] shows that this equation is generically uniquely solvable for the elements $\{\Psi_{ij}^*\}$, although the algebra can become a bit tedious when m and/or p are greater than 1. However, in the single-input, single-output case, it's possible to work-out the explicit representation

$$
\Psi_{\omega^*, f^*} =
\begin{bmatrix}
\mathcal{R}_1^* u_0^* & 0 & 0 & \cdots \\
\mathcal{R}_2^* u_0^* + \mathcal{R}_1^* u_1^* & \mathcal{R}_1^* u_0^* & 0 & \cdots \\
\mathcal{R}_3^* u_0^* + \mathcal{R}_2^* u_1^* + \mathcal{R}_1^* u_2^* & \mathcal{R}_2^* u_0^* + \mathcal{R}_1^* u_1^* & \mathcal{R}_1^* u_0^* & \cdots \\
\vdots & \vdots & \vdots & \cdots
\end{bmatrix}.
$$

Example: The Natural Numbers (cont'd.)

In our continuing example, suppose the basal metabolism f^* is perturbed to the new metabolism

$$
f = \{1, 2, 2, 2, 3, 3, 4, 4, \ldots\} = \{A_1, A_2, A_3, \ldots\},
$$

i.e., there is only a change in the 2nd element. Under this metabolism, the cellular output is now

$$
\gamma = f(\omega^*) = (1, 3, 4, 4, 5, 6, 7, \ldots).
$$

Thus, the metabolic change results in a change of output from the design output $\gamma^* =$ the natural numbers to the closely related sequence

γ, which differs from γ^* only in the 2nd and 3rd entries. Our question is the manner in which the repair mechanism acts to stabilize the cell in the face of this change in the basal metabolism.

To address this issue, we appeal to the Metabolic Repair Theorem and compute the matrix Ψ_{ω^*, f^*}, using the representation for single-input, single-output systems given above. It's easily seen that for this case we have

$$\Psi_{\omega^*, f^*} = \begin{pmatrix} 1 & 0 & 0 & \cdots \\ 0 & 1 & 0 & \cdots \\ 0 & 0 & 1 & \cdots \\ \vdots & \vdots & \vdots & \ddots \end{pmatrix} = \text{identity}.$$

Thus according to the Metabolic Repair Theorem we are in the case where the cell will lock-on to the new metabolism f. In fact, by a slightly more technical argument it can be shown that this will *always* be the situation for single-input, single-output systems. Such systems for which $m = p = 1$ will always have $\Psi_{\omega^*, f^*} = \text{identity}$, implying that metabolic disturbances for such cells can never be repaired; the cell can only, in effect, start tracking whatever metabolic "noise" may be introduced into the situation. However, this need not be the case for cells having more than a single-input and/or a single-output, as is shown by example in [7]. Having now considered the situation when there is a metabolic disturbance, let's turn to a consideration of cellular repair in the face of changes to the design environment ω^*.

Case II: A fluctuating environment ω with a fixed basal metabolism f^* and a fixed genetic map P_{f^*}.

In this situation, we have a change of environment $\omega^* \to \omega$, and we want to determine those environments ω for which

$$P_{f^*}(f^*(\omega^*)) = P_{f^*}(f^*(\omega)) \quad (= f^*)$$

implies

$$f^*(\omega^*) = f^*(\omega).$$

In other words, we want to know when the map P_{f^*} is one-to-one.

From the matrix representation of the map P_{f^*}, it's a simple matter to establish the

ENVIRONMENTAL REPAIR THEOREM. *If $m = 1$ and rank $\mathcal{R}_1^* = p$, all environmental disturbances ω that are repaired by the cellular genetic mechanism are given by $\omega = \omega^* + \ker f^*$.*

On the other hand, if $m > 1$ and/or rank $\mathcal{R}_1^* = r < p$, then the only types of environmental changes that can be repaired are those of the form $\omega = x + \omega^*$, where x is any solution of the equation $f^*(x) = \hat{\gamma}$, with $\hat{\gamma} \in \ker \mathcal{R}_1^*$.

The Metabolism and Environmental Repair Theorems cover the cases of interest insofar as determination of what kinds of disturbances to the cell can be neutralized by the genetic machinery. Now let's briefly consider the replication mechanism, and ask under what conditions "mutations" in the genetic map P_{f^*} can be preserved in the cellular genotype by means of the replication operation.

7. Linear (M, R)-Systems: Replication

The cellular replication map

$$\beta_f \colon H(\Omega, \Gamma) \to H(\Gamma, H(\Omega, \Gamma)),$$

can be formally considered in much the same fashion as just discussed for the repair mechanism P_f. However, since the functional role of replication is quite different from that of repair, a number of interesting questions arise that are absent in the case of repair, questions involving mutation, adaptation, Lamarckian inheritance and so forth. I'll consider some of these matters in a moment, but first let's look at the problem of formal realization of the replication map β_f.

By linearity, the action of β_f can be represented as

$$c_\sigma = \sum_{i=0}^{\sigma-1} \mathcal{U}_{\sigma-i} e_i,$$

for an appropriate set of matrices $\{\mathcal{U}_i\}$, where the input $e_i = \mathcal{S}(A_i)$ and the output $c_i = \mathcal{R}_i$. Here $\mathcal{S}(A)$ is the "stacking" operator, whose action is to stack the columns of A into a column vector of pm components. Arguing just as for the repair map, we find that the elements $\{\mathcal{U}_i\}$ must have the form

$$\mathcal{U}_j = [\mathcal{C}_{j1} | \mathcal{C}_{j2} | \cdots | \mathcal{C}_{j,mp}], \qquad j = 1, 2, \ldots,$$

where each $\mathcal{C}_{jr} \in R^{p \times mp}$. Note also that here we have again introduced an additional time-scale σ for operation of the cellular replication subsystem, which will generally be at a slower scale than either the metabolism or the repair operations. Also observe the added degree of complexity in the operators defining the replication map β_f, since each of the elements $\mathcal{U}_i$ requires $O(m^2 p^3)$ numbers, a factor pm greater than for the repair operators $\mathcal{R}_i$, and a factor $p^2 m$ greater than for the metabolic operators A_i. Further, using the same arguments as for P_f, it can be easily established that if f has a finite-dimensional realization, so does β_f and $\dim \beta_f \leq \dim f$.

Example: The Natural Numbers (cont'd.)

Our standard example has

$$f^* = \{1,1,2,2,3,3,\ldots\} = \{A_1^*, A_2^*, A_3^*,\ldots\},$$
$$P_{f^*} = \{1,-1,1,-1,\ldots\} = \{\mathcal{R}_1^*, \mathcal{R}_2^*, \mathcal{R}_3^*,\ldots\}.$$

After a bit of algebra, we then find that

$$\beta_{f^*} = \{1,-2,1,0,0,\ldots\} = \{\mathcal{U}_1^*, \mathcal{U}_2^*, \mathcal{U}_3^*,\ldots\}.$$

Thus, only the first three terms $\mathcal{U}_1^*, \mathcal{U}_2^*$ and $\mathcal{U}_3^*$ are non-zero. Applying Ho's Algorithm to the external description β_{f^*}, we quickly find the internal dynamics canonically realizing β_{f^*} as

$$q_{\sigma+1}^* = \begin{bmatrix} 0 & 0 & 0 \\ 1 & 0 & 0 \\ 0 & 1 & 0 \end{bmatrix} q_\sigma^* + \begin{bmatrix} 1 \\ 0 \\ 0 \end{bmatrix} e_\sigma^*, \qquad q_0^* = 0,$$
$$c_\sigma^* = [1 \ -2 \ 1] q_\sigma^*, \qquad \sigma = 1,2,\ldots.$$

There are two basic questions that we would like to address in the context of replication:

1) When can changes in the design environment ω^* and/or in the basal metabolism f^* result in changes to the replication map β^*? This is what I term the Lamarckian Problem, as it forms an abstract version of the classical Lamarckian question of whether environmental and/or phenotypic changes can be transmitted to descendants.

2) Under what circumstances are changes in the genetic map P_f preserved by corresponding changes in the replication map β_f? I term this the Mutation Problem, as it abstractly expresses the usual issue of when external perturbations to the cellular genotype are inherited by the cell's "offspring."

For the sake of space, let me treat here only the Lamarckian Problem in the case of environmental changes. For the interested reader, a more detailed study of the other questions can be found in [7].

From the diagram

$$\Omega \xrightarrow{f^*} \Gamma \xrightarrow{P_{f^*}} H(\Omega, \Gamma) \xrightarrow{\beta_{f^*}} H(\Gamma, H(\Omega, \Gamma)),$$

it's evident that

$$P_{f^*}(f^*(\omega^*)) = f^* = [\beta_{f^*}(f^*)] \, f^*(\omega^*).$$

Suppose we have a change of environment $\omega^* \to \omega$. This results in a change of output

$$\gamma^* = f^*(\omega^*) \to f^*(\omega) = \gamma.$$

Assume that $P_{f^*}(\gamma) = P_{f^*}(\gamma^*) = f^*$, i.e., the repair mechanism is capable of compensating for the environmental change. Then we have

$$(\beta_{f^*} \circ P_{f^*})(\gamma^*) = (\beta_{f^*} \circ P_{f^*})(\gamma) = P_{f^*},$$

implying that the replication operation is unaffected by the environmental change. That is, Lamarckian-type changes in β_{f^*} cannot occur under any type of environmental change that can be compensated for by the cellular repair operation P_{f^*}. The Environmental Change Theorem characterizes just what sorts of changes fall into this category.

Now let's move away from the theoretical aspects of the (M, R)-systems, and consider some ways in which they can be used to represent problems of applied interest.

8. Applications of (M, R)-Systems in Biology, Economics and Manufacturing

The abstract development of the (M, R)-systems given above makes it evident that these systems can be used to mathematically represent any sort of process in which there is an underlying metabolic aspect representable by some kind of input/output behavior. From the given I/O behavior, I've shown how to use it to construct both the system's internal repair and replication dynamics by straightforward, "natural" mathematical operations which, in the linear case considered here, comes down to a relentless use of the realization algorithm in the right formal setting. In the next section I'll take up various extensions, generalizations and possible limitations of this entire picture, including the matter of nonlinearity, but for now I want to give some indication of just how the formal (M, R)-framework can be used in an applied mode to shed light on a few of the types of living systems alluded to at the beginning of this essay. Unfortunately, space constraints prevent a detailed examination of these application areas, but I hope that the general outline provided here can be used as the starting point for a more thorough exploration of the implications of the (M, R)-formalism for the treatment of the important questions considered below. Now let's look at the application areas in turn.

• *Cellular Biology*—by construction, the (M, R)-systems can be employed to capture the basic *functional* activities of a living cell. However, it's been my experience that most biologists, even those of a theoretical stripe, are deeply suspicious of any kind of theoretical framework that isn't firmly rooted in the physico-chemical nature of cellular activity. According to this narrow view, if a cellular modeling framework doesn't talk about the specific chemical activities going on in the cell, then that framework can tell you nothing of interest about **real** cells and, hence, is unworthy of mention in polite biological conversation. As already noted, this is a modeling attitude that locks one into a consideration of material and efficient cause, claiming at least implicitly that formal and final cause play no role whatsoever in explaining the questions of interest in real cells. In order to make contact with this kind of mainstream attitude it's of interest to see the degree to which the (M, R)-paradigm can be used to represent some of the physico-chemical aspects of cellular operation.

At the outset it should be clear that just as Newtonian mechanics gives one no insight into the actual physical morphology of planetary bodies, the (M, R)-systems are inherently incapable of addressing morphological questions pertaining to the actual geometric structures which a living cell may assume. Thus, none of the usual questions relating to the material and efficient causes in embryology, cellular differentiation, morphogenesis and so forth come within the purview of our (M, R)-modeling framework. However, this is not to say that we cannot address other types of questions of a *non-structural,* yet still physico-chemical nature. The essence of making this sort of contact with real cellular operation lies in a meaningful identification of the abstract input and output elements with actual chemical compounds.

In very crude terms, the business of every living cell is the manufacture of proteins of various sorts, depending upon the specific instructions coded into the cellular DNA. Thus, to use our (M, R)-framework to represent this process, we must identify several major elements:

1) The types of chemical compounds available to the given cell. Finite sequences of amounts of these compounds then form the elements

of our abstract set Ω. Of course, such an identification means we must determine just how much of each elementary compound will be available to the cellular processing machinery at each time instant of the process.

2) The types of chemical compounds that the particular type of cell under study is designed to produce. Of course cells produce many different kinds of compounds, but the most important are the various proteins that the cell manufactures in order to carry out its assigned function. The sequences of proteins and intermediate products that the cell produces can then serve as the elements of the output set Γ.

3) Once the design environment $\omega^* \in \Omega$ and design output $\gamma^* \in \Gamma$ have been set, the basal metabolism f^* is then fixed by the relationship $f^*(\omega^*) = \gamma^*$. In the linear situation we have been considering here, it can be shown that if it is assumed beforehand that the internal dynamics can be realized by a finite-dimensional system, then f^* will be uniquely determined for almost every choice of input/output pair (ω^*, γ^*). Nonlinear extensions will be considered a bit later.

Specification of the elements (1)–(3) fixes only the cellular metabolism; however, as has been demonstrated above, this is all that's necessary in the (M, R)-framework to then determine the entire genetic machinery of the cell, as well. So the job of translating the abstract metabolism-repair formalism into an actual model of a single living cell comes down to determining a scheme for coding the real cellular chemical inputs into the abstract elements of ω, as well as development of a similar coding scheme for the metabolites into the abstract elements of γ. I suspect that this is a far more difficult task than it looks on the surface, as real cells are capable of lots of different activities that will likely make it tricky to fit them into such a neat package as a finite-dimensional vector space. Thus, many simplifications will have to be made and many aspects of real cellular behavior will have to be ignored. Only cellular biologists familiar with the intricacies of actual cellular activity will be in a position to decide just what should and shouldn't be included in such an exercise, but it certainly doesn't seem to me to be an insurmountable task, and I don't see any *a priori* reasons why the necessary simplifications need be any more outlandish than the kinds of physical fictions that we already take for granted, such as point particles, centers of mass and point charges. Not being a cellular biologist, I'll happily leave this chore to those inclined to pursue it and move on to an application area where *everyone* is an expert—economics.

• *Input/Output Economics*—despite the fact that the underlying motivation for the development of the (M, R)-systems was to capture the essential functional activities of a living cell, the water is somewhat murky insofar as actually using the abstract formalism to attack real cellular questions, partially due to the excessive pre-occupation of cellular biologists with questions pertaining to material and efficient cause. Amusingly enough, such obstacles vanish when we turn our spotlight in another direction and examine the possibility of using the (M, R)-paradigm in economics. In contrast to biology, in economics there are virtually no questions of any interest involving material cause, and almost everything that looks to be of even faint economic interest revolves about matters of formal and final cause—just the territory for the (M, R)-framework.

Much of the modern work in theoretical economics is filled with terms such as "evolutionary economics," "adaptive economic structures," "self-organizing economies," and so forth. These are evocative terms, suggesting a strong need for an underlying biological metaphor suitable for modeling modern economic processes. Space forbids a detailed consideration of this metaphor here, so instead I'll look at a far more mainstream view of economic processes, the classical input/output picture pioneered by Leontieff, and show how this paradigm can be substantially extended by invoking the (M, R)-formalism.

To illustrate the basic ideas involved in using the (M, R)-formalism for Leontieff-type economies, consider the very simplified Leontieff system in which the system dynamics are

$$x_{t+1} = Fx_t + Gu_t, \qquad x_0 = x_0,$$

where the ith component of the vector $x_t \in R^n$ represents the level of the ith product in the economic complex at time t, with the production matrix F having the form

$$F = \begin{pmatrix} 0 & 0 & \cdots & 0 & a_1 \\ a_2 & 0 & \cdots & 0 & 0 \\ 0 & a_3 & \cdots & 0 & 0 \\ \vdots & \vdots & \vdots & \vdots & \vdots \\ 0 & 0 & \cdots & a_n & 0 \end{pmatrix}, \qquad a_i \geq 0.$$

The ith element of the input vector u_t represents the labor input to the ith product at time t, with the matrix G having the form

$$G = \text{diag}\,(g_1, g_2, \ldots, g_n), \qquad g_i > 0.$$

Further, assume that the products $i = 1, 2, \ldots, n - 1$ are intermediate products, so that the level of finished product in the economy is measured by the nth component of x_t. Thus, the economy's output at time t is

$$y_t = x_t^{(n)} = H x_t = (0 \ 0 \ \cdots \ 0 \ 1) x_t.$$

The above situation is already cast in exactly the form needed for the (M, R)-formalism. The "environment" ω is specified by giving the sequence of labor inputs $\{u_0, u_1, \ldots, u_N\}$, while the "cellular output" of the economy is the level of finished product $x_t^{(n)}$, i.e., $\gamma = \{x_1^{(n)}, x_2^{(n)}, \ldots, \}$. As soon as we specify the "basal" levels of these quantities ω^* and γ^*, then we are in position to employ the entire (M, R)-framework discussed earlier in order to construct the repair and replication components of this economic system. Furthermore, since the actual internal dynamics of the system are given, we can directly calculate the basal metabolism as $f^* = \{A_1^*, A_2^*, \ldots\}$, where the elements $A_i^* = H^*(F^*)^{i-1} G^*$. Once these "non-Leontieffian" components of the system are in place, we can begin to ask the by now familiar questions about what kinds of changes in labor supply and/or technological coefficients can be compensated for by the system, what kinds of "mutations" in the technological coefficients will be permanently imprinted upon the system and so on. We leave it to the reader to formulate these problems in more precise terms, and now turn our attention to our final application area, manufacturing.

● *Manufacturing Systems*—a much more specific area of potential application of the (M, R)-framework is for the modeling of the so-called "Factory of the Future," in which the manufacturing system is designed to be self-organizing, self-repairing, and even in some cases, self-regenerating. Let me sketch the basic ideas, referring the reader to the papers [8–9] for further details.

Every manufacturing operation can be thought of as a metabolic process in which a set of raw materials is transformed by labor, money and machines into an end product. At first sight, this sounds suspiciously like another case of material and efficient cause; but further examination shows that the real issues under discussion in manufacturing circles today involve matters of information manipulation, i.e., formal cause. The real concern is with just exactly how the processing of the raw materials into finished products should be organized to satisfy various constraints relating to flexibility, reliability, quality, adaptability, etc. of the manufacturing operation. These are issues of

formal and final cause and as such fall smack-dab in the middle of the
(M, R)-formalism.

Assume that Ω represents the set of sequences of available raw
material inputs that can be used by the manufacturing enterprise, with
Γ being the corresponding set of sequences of finished products that the
factory is capable of producing. Just as in the biological case, to make
contact with real manufacturing processes we would have to find a good
way of encoding these items of the real-world into the abstract elements
of the sets Ω and Γ in order to invoke the (M, R)-paradigm.

For example, if the factory is supposed to produce automobiles
of a certain type, then this means that the real-world inputs such as
the amount of steel, glass, rubber, plastic, hours of skilled labor, hours
of unskilled labor, number of machine-hours for welding, drilling, etc.
would all have to be encoded into the elements of the input u_t for
each time t. This encoding would then specify the input sequence
$\omega = \{u_0, u_1, \ldots, u_N\}$. A similar encoding would have to be developed
for the output elements $y_1, y_2, \ldots$, expressing the sequence of partial
products of the overall auto manufacturing operation. This encoding
would then determine the output sequence $\gamma \in \Gamma$. Then a particular
type of automobile would be specified by giving a *particular* input and
output sequence ω^* and γ^*. At this point, we are in a position to
employ the (M, R)-setup to determine the corresponding repair and
replication subsystems for this manufacturing enterprise.

Again, just as with the problem of using the abstract (M, R)-
framework in cellular biology, there are a number of non-trivial questions that arise in connection with actually translating the foregoing
ideas into workable schemes for real-world operations. But the main
point is that the framework exists, and the tasks that need to be carried
out to make contact with real-world problems are fairly clear and explicit. All that's needed is the will and time to see the exercise through
to conclusion, exactly the same sort of requirements needed to use the
Newtonian formalism to study the behavior of the mechanical processes
of physics. The only difference is that the "Newton of Manufacturing"
has not yet made his appearance, but the necessary formalism has.
Let's now consider a spectrum of extensions and generalizations that
may be needed to effectively make use of the basic set-up in a given
real-world situation.

9. Networks, Nonlinearities and Evolution

The development of the preceding sections has been confined to the
simplest of all possible settings—a single cell operating according to

linear dynamics. It might be argued that these are restrictions ruling out consideration of any real cellular processes, and in a weak moment I might be inclined to agree. Nonetheless, from a theoretical point of view it's necessary to walk before we can fly, and the simple setting considered here is useful for at least two reasons: i) it demonstrates the feasibility of using the (M, R)-systems as a theoretical generalization of the classical Newtonian framework, and ii) it points the way to the new kinds of questions that will remain important even under more general circumstances. But, interesting as the single-cell, linear case may be, we must consider how to "soup-up" this situation if we are to make contact with the processes of real-life. Consequently, let me here outline some of the major extensions needed and give some indication of how I think they can be carried out. By and large, these extensions fall into three broad categories, each involving a weakening of one of the underlying assumptions in the models considered above.

The first condition that needs relaxing is the confinement to a single-cell system. It's necessary to consider *networks* of such cells coupled together in various ways. Just as Newtonian mechanics is pretty trivial for single-particle systems, if we want to use the (M, R)-systems as a modeling paradigm for real organisms, consideration of networks is an essential step that must be taken.

The second constraint we need to address is the matter of linearity. Experimental evidence is rather clear on the point that biological cells do not, in general, act in a linear manner when transforming one set of chemicals into another. For example, there are saturation limits at which addition of further environmental inputs produces no additional output. Thus, it will be necessary to see how and to what degree it's possible to push through the constructions given above for cellular metabolisms displaying various types of nonlinear effects.

Finally, the cellular systems considered here have no real final causation built into them. In other words, we have imposed no kind of optimality criterion by which the cell can decide whether or not certain types of modified metabolisms or new genetic structures are to be favored in the battle for survival. Thus if we want to use the (M, R)-structure as a framework to study evolutionary processes, we will have to superimpose some kind of selection criterion on the genetic makeups of the cells so that natural selection can take its course. It should be evident that these sorts of evolutionary considerations must be added to the general picture after dealing with the networking problem discussed a moment ago. With the above considerations in mind, let me take a longer look at each of these extensions.

• *Cellular Networks*—the most straightforward way in which to think about putting together a collection of individual cells into a network is to assume a situation in which at least one cell of the network receives part of its input from the environment, with at least one cell transmitting part of its output to the environment. All cells not interacting directly with the environment then receive their inputs as the outputs of other cells, and transmit their outputs to be used as the input to other cells. We assume that each cell must have at least one input and one output, and that proper metabolic functioning of the cell requires the cell to receive all of its inputs. This arrangement takes care of the metabolic activity of the network, but we must also make some sort of assumptions about the connectivity pattern involved with the individual genetic components of the network elements. The simplest is to assume that the repair component of each cell satisfies the following conditions: i) it receives at least one environmental output as part of its input, and ii) it must receive all of its inputs in order to function. A typical example of such a network is depicted in Fig. 1, where the boxes labeled "M" represent the cell's metabolic component, while the ovals labeled "R" are the corresponding repair/replication components.

In this network, observe that if cell 1 fails, then so will cells 2–5, all of whose inputs ultimately depend upon the proper functioning of cell 1. We could call any such cell whose failure entails the collapse of the entire network, a *central component* of the network. Now note that by the hypotheses made above concerning the repair components, any cell whose repair component receives the output of cell 1 cannot be built back into the network by the repair component. Such a cell can be termed *non-reestablishable.* Thus, cell 2 is non-reestablishable, while cell 5 is reestablishable. These elementary ideas already lead to the following important result of Rosen's [1] about (M, R)-networks:

(M, R)-NETWORK THEOREM. *Every finite (M, R)-network contains at least one non-reestablishable cell.*

COROLLARY. *If an (M, R)-network contains only a single non-reestablishable cell, then that cell must be a central component of the network.*

The Network Theorem and its Corollary can be proved by a simple inductive argument, and show that every cellular network must contain elements that cannot be built back into the system should they fail. Further, if there are only a small number of such cells, then the failure of these cells is likely to bring down the entire network. This last is a point to ponder in connection with various types of social policies

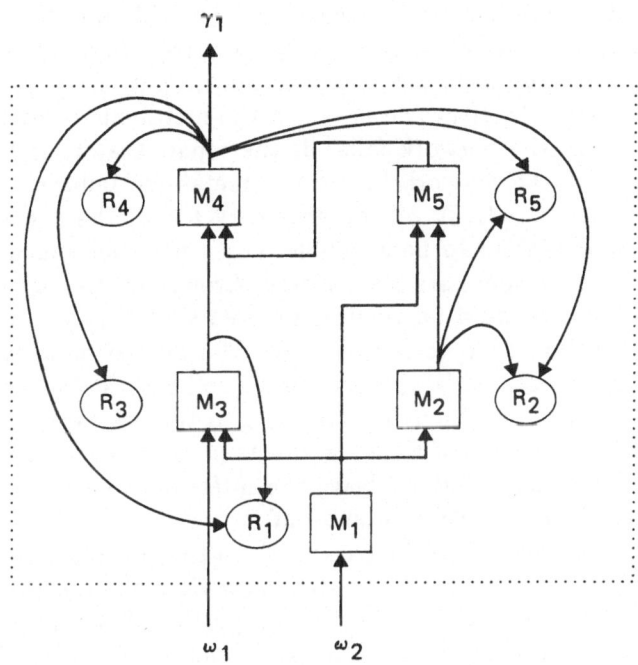

Figure 1. A Typical Cellular Network

aimed at propping-up failing enterprises. The Network Theorem says that such propping-up cannot possibly be universally successful and, what's worse, the more successful it is, the greater is the likelihood of creating a system in which the failure of a very small number of enterprises will result in the collapse of the entire edifice.

Of course, the Network Theorem above depends on the particular assumptions made here about the need for the metabolic and repair subsystems to each have all their inputs in order to function, as well as upon the hypothesis that each repair subsystem must receive at least one environmental output as one of its inputs. In any real cellular network, these conditions may or may not hold and, if not, it will be necessary to see to what degree the kind of result obtained above can be carried over under the new conditions.

   In addition to problems involving connectivity patterns, there are important time-lag effects that enter when we begin discussing networks. Besides the internal time-scales associated with the metabolic, repair and replication subsystems of each individual cell, when we start connecting such cells together we quickly encounter another type of time-lag, the time needed for information to pass from one cell to another. Imagine the situation in which the metabolic component of one of the cells in the network fails. If the repair system is to build this cell back into the network before the metabolic failure can be transmitted to other cells, it will be necessary for all the necessary inputs to the repair system to be available on a time-scale much faster than the metabolic outputs are transmitted throughout the network. If not, then the failure cannot be repaired before its effect is felt on other cells, thereby causing them to fail as well. The degree to which this local failure can be isolated is directly related to how fast the corresponding "raw materials" can be made available to the relevant repair system. There are other sorts of time-lag effects of a metabolic type that also enter into networks, but I'll leave the interested reader to consult the discussion given in [1] for an account.

   Once we have made the passage from the consideration of a single cell to a network of such cells, a whole new vista of opportunity presents itself for the study of the collective properties of the network and the way in which the collective properties emerge from the behavior of the individual cells. For example, there are the kinds of survivability issues that the Network Theorem only begins to touch. Or in another direction, problems of adaptation and evolution of the cells which we'll take up below. Such evolutionary questions are also closely related to the problem of cellular differentiation, in which the metabolic activity of some cells changes in a manner so as to make the behavior of the overall network more efficient, in some sense. The (M, R)-framework offers a paradigm for the theoretical consideration of all these issues, as well as much, much more. Let me now move away from questions surrounding networks and groups of cells, and take a longer look at the problem of nonlinearity.

   • *Nonlinearities*—the technical details about (M, R)-systems given in the preceding sections, skimpy as they are, all rely upon the assumption that the cellular metabolic, repair and replication outputs are linear functions of the inputs. From a mathematical standpoint, this is quite convenient as there is by now a very well-developed literature on linear systems which I have been able to tap into in order to provide a foundation for a technical treatment of the processes of repair and reproduction. However, there's good reason to look at this linearity hypothesis with a fishy eye, since the experimental evidence on cellular input/output behavior (what little there is of it) doesn't appear to strongly support the case for linearity. In other settings, however, as indicated in the Leontieff-type economy example, the linearity hypothesis might be more easy to swallow, but the truth of the matter is that for the (M, R)-framework to provide a solid basis for modeling living processes, it's necessary to somehow extend the linear results to a wider setting.

The results given earlier under the general rubric "Repair Theorems" relied mainly upon the structure of the kernel of certain linear operators, while the algorithm associated with actually constructing the cellular metabolic and genetic dynamics came out of the Ho procedure for constructing canonical realizations of linear input/output behaviors. These are both classical ideas in linear algebra. If the (M, R)-setup is to be extended to deal with interesting classes of nonlinear behaviors, it will be necessary to find suitable nonlinear extensions of these basic linear concepts.

Fortunately, the past decade or so has seen an explosion of interest in the problems of *nonlinear* system theory, and most of the machinery needed to extend the underlying vector-space setting, as well as the Ho Realization Algorithm to broad classes of nonlinear systems is now in place [10–11]. Of course, the mathematical tools needed for these extensions are quite a bit more sophisticated, in general, than the simple tools of linear algebra, although a number of important extensions to systems with some kind of linear structure (e.g., bilinear or multilinear I/O maps) can be pushed through using the linear apparatus. I have given a more complete account of how to make some of these extensions in a forthcoming technical paper [7], so I won't go into details here other than to state that there is no theoretical obstacle whatsoever to carrying out the same kind of analysis done here but in the setting of nonlinear cellular behaviors. The surface form of the results will be somewhat more complicated, requiring some of the ideas and terminology from modern abstract algebra and differential geometry to properly state, but the underlying concepts and (probably) the results will be very similar to the linear case.

• *Adaptation and Evolution*—the single-cell situation treated here assumes that the cell has only a single purpose: to fulfill the prescription laid down in its design metabolism. Thus, there is no explicit criterion of performance, as the entire genetic apparatus is set-up solely to insure that the cell sticks to the basal metabolism, if at all possible.

Such a narrow, purely survival-oriented objective is called into question when we move to the setting of cellular networks in which the cell is not just in business for itself, but also to serve the needs of the other cells. Thus, in this kind of setting it makes sense to superimpose some sort of evolutionary fitness criterion upon the basic survival condition, thereby measuring the degree to which the cell is "fit" to operate within the constraints of its "generalized" environment, consisting now of not only the external physical environment, but also the other cells in the network. As soon as such a selection criterion is imposed, we then have all the ingredients necessary for an evolutionary process, since we have already mentioned ways in which the genetic subsystem can experience various sorts of "mutations."

It is by now folk wisdom in the control engineering business that choice of a performance criterion is usually the trickiest aspect of putting together the pieces of a good optimal control model for a given situation. The situation is no easier in biology, where the arguments still rage hot and heavy over competing positions on the matter of selection criteria for evolutionary processes. In fact, in biology the situation is, if anything, worse, since there is not even a consensus as to the level at which the selection acts, with rather vocal arguments being made for action at every level from the gene itself up to the complete organism. Consequently, I'm afraid there is no divine mathematical insight that can be offered as to how to choose an appropriate selection criterion to use with the (M, R)-framework we've introduced here. About all that can be said is that once such a choice has been made, then a network of cells put together according to the (M, R)-prescription will provide a natural theoretical basis for the study of adaptive and evolutionary processes.

References

[1] Rosen, R., "Some Relational Cell Models: The Metabolism-Repair Systems," in *Foundations of Mathematical Biology*, R. Rosen, ed., Volume 2, Academic Press, New York, 1972.

[2] Rosen, R., "A Relational Theory of Biological Systems," *Bull. Math. Biophysics*, 21 (1959), 109–128.

[3] Rosen, R., "A Note on Absract Relational Biologies," *Bull. Math. Biophysics*, 24 (1962), 31–38.

[4] Casti, J., *Linear Dynamical Systems*, Academic Press, New York, 1987.

[5] Brockett, R., *Finite-Dimensional Linear Systems*, Wiley, New York, 1970.

[6] Fortmann, T. and C. Hitz, *An Introduction to Linear Control Systems*, Dekker, New York, 1977.

[7] Casti, J., "The Theory of Metabolism-Repair Systems," *Appl. Math. & Comp.*, to appear 1989.

[8] Casti, J., "(M, R)-Systems as a Framework for Modeling Structural Change in a Global Industry," *J. Social & Biological Structures*, to appear 1989.

[9] Casti, J., "Metaphors for Manufacturing: What Could it be Like to be a Manufacturing System?" *Tech. Forecasting & Social Change*, 29 (1986), 241–270.

[10] Isidori, A., *Nonlinear Control Systems: An Introduction*, Springer, Heidelberg, 1985.

[11] Casti, J., *Nonlinear System Theory*, Academic Press, New York, 1985.

Some Thoughts on Modelling

JAN C. WILLEMS

Abstract

The purpose of this article is to explain in simple terms our way of approaching mathematical models, in particular models for dynamical systems. We will put forward some ideas on the following topics:
  1. How should one view models as mathematical entities?
  2. How can models be deduced from observed data?

I. General Models

1. Introduction

Is the dictum that **modelling is more of an art than a science** *a deep truth or a self-serving and self-fulfilling platitude? Does it make sense to try to find some unity, some classification, some distinguishing features in the multitude of model classes that have been proposed and studied, for example, in dynamical system theory?* Our present research attempts to achieve this. We believe that by viewing models in terms of their behavior and by incorporating latent variables into the modelling outlook at the very beginning, one can obtain a theory which is both general and consistent and which can even point to and yield new algorithms in such areas as system identification.

What really constitutes a mathematical model? What does it tell us? What is its mathematical nature? Mind you, we are not asking a philosophical question; we will not engage in an erudite discourse about the relation between reality and its mathematical description. Neither are we going to elucidate the methodology involved in actually deriving, setting up, and postulating mathematical models. What we are asking is a simple question: When we have accepted a mathematical expression, a formula, as an adequate description of an object of study, what mathematical structure have we really obtained? Our question is addressed to the engineer, the mathematician, the economist who, on the one hand, having learned much of what there is to know about sets, maps, solving equations, and so forth, and who, on the other hand, having seen all sorts of models of mechanical gadgets, electrical devices, and economic processes, must have acquired a good intuitive feeling for what mathematical modelling is all about. Has this led to a synthesis, an abstract view, a conception of the notion of a mathematical model?

Our questions are, unfortunately, rhetorical, since very little attention
has been paid to such issues in education.

2. Mathematical Models

Assume that we start with a phenomenon which, for one purpose or
another, we have decided to model. We postulate that this phenomenon
can be described by *attributes,* certain variables which for our aims
describe the outcomes of the phenomenon. The set of all values of the
phenomenon which can conceivably occur (before we engage in taking
a closer look, i.e., modelling, the phenomenon), forms what we will call
the *universum.* A mathematical model claims that some values of the
attributes are believed to be possible, while others are not. Hence a
mathematical model is simply a subset of the universum. Formally:

DEFINITION. *A mathematical model is a pair* $(\mathfrak{U}, \mathfrak{B})$ *with* $\mathfrak{U}$ *the universum of conceivable attributes and* $\mathfrak{B} \subseteq \mathfrak{U}$ *the behavior of the model.*

We will now illustrate this definition by means of an elementary example.

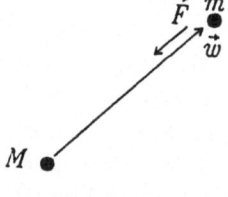

Figure 1

The inverse square law: Three hundred years ago, Sir Isaac Newton
discovered (better: deduced from Kepler's laws since, as he put it:
hypotheses non fingo) that masses attract each other according to the
inverse square law. Let us formalize what this says about the relation
between the force $\vec{F}$ and the position vector $\vec{w}$ of the mass m shown
in Figure 1. We assume that the other mass M is located at the
origin of $\mathbf{R}^3$. The universum $\mathfrak{U}$ consists of all conceivable force/position
vectors, yielding $\mathfrak{U} = \mathbf{R}^3 \times \mathbf{R}^3$. After Newton explained to us what was
happening, we knew more:

$$\mathfrak{B} = \left\{ (\vec{F}, \vec{w}) \in \mathbf{R}^3 \times \mathbf{R}^3 \,\middle|\, \vec{F} = -k \frac{mM\vec{w}}{\|\vec{w}\|^3} \right\},$$

with the gravitational constant $k = 6.67 \times 10^{-8} \mathrm{cm}^3/\mathrm{gram} \cdot \mathrm{sec}^2$. Note
that $\mathfrak{B}$ has three degrees of freedom—down three from the six degrees
of freedom in $\mathfrak{U}$.

3. Behavioral Equations

Let us now take a closer look at some possible specifications of a mathematical model. The most elementary specification, the one which will come to mind almost immediately, is the specification by means of equations. In physical systems these equations may be a combination of constitutive equations, field equations, conservation laws, etc. The behavior of a model consists of those elements of the universum which satisfy these equations. Formally, we are given two maps (b_1, b_2) from the universum $\mathfrak{U}$ into a space $\mathfrak{E}$, called the *equation space*. This yields the relation

$$b_1(u) = b_2(u).$$

We will call this the defining *behavioral equation*. It defines the model $(\mathfrak{U}, \mathfrak{B})$ with $\mathfrak{B} = \{u \in \mathfrak{U} \mid b_1(u) = b_2(u)\}$. The example given above provides a clear illustration of a mathematical model described by a behavioral equation, the inverse square law.

The defining behavioral equations uniquely specify the model, the behavior $\mathfrak{B}$, but the converse is clearly not true. This point is often overlooked, for example, in model fitting where *equation error* is minimized. Indeed, since the same model can be described by many different sets of equations, this tends to introduce an undesirable element of arbitrariness into these procedures.

4. Latent Variables

The primary purpose of a mathematical model is to describe the behavior of the attributes. However, as can be seen from examples, the prescription of the behavior can occur in many different forms. Often, in fact, such prescriptions will involve latent variables, auxiliary variables which facilitate describing the behavior of the attributes.

DEFINITION. *A mathematical model with latent variables consists of a triple $(\mathfrak{U}, \mathfrak{L}, \mathfrak{B}^e)$, with $\mathfrak{U}$ the **universum**, $\mathfrak{L}$ the set of **latent variables**, and $\mathfrak{B}^e$ a subset of $\mathfrak{U} \times \mathfrak{L}$, called the **extended behavior**. The resulting **intrinsic behavior** of the attributes is given by $\mathfrak{B} = \{u \in \mathfrak{U} \mid \exists \ell \in \mathfrak{L} \text{ such that } (u, \ell) \in \mathfrak{B}^e\}$. In other words, $\mathfrak{B} = P_u \mathfrak{B}^e$ with $P_u : \mathfrak{U} \times \mathfrak{L} \to \mathfrak{U}$, the projection $(u, \ell) \overset{P_u}{\longmapsto} u$.*

Latent variables may sound abstract. Let us illustrate the idea by means of a number of word examples:

• When writing down a model of an electrical circuit, an electrical engineer will need to introduce the currents through and the voltages across the internal branches of the circuit in order to express the constraints imposed by the constitutive laws of the elements and Kirchhoff's current and voltage laws. These internal voltages and currents can be viewed as latent variables.

• When setting up a model for the dynamics of the positions of the moving parts of a machine, a mechanical engineer may find it convenient to introduce the momenta as latent variables.

• When formulating the laws of thermodynamics, it is useful to introduce the internal energy and the entropy. These are latent variables.

• When postulating a relation for the time evolution of the demand and supply of a scarce resource, an economist may want to introduce the price of this resource as an auxiliary variable. The price can be considered as a latent variable.

• When explaining the scores on tests, a psychologist will find it useful to consider intelligence as a latent variable.

• When axiomatizing the nature of the memory of a dynamical system or when studying its stability, a mathematician will be led to write the equations in state form. The state becomes a latent variable.

Let us work out the supply/demand example in somewhat more detail.

Supply and demand: An economist is trying to figure out how much will be produced and consumed of a certain economic resource. Since (s)he believes in market equilibrium for the effective functioning of the economy, (s)he postulates *a priori* that the supply S will equal the demand D. This yields the universum $\mathfrak{U} = \{(S, D) \in \mathbf{R}_+^2 \mid S = D\}$. How can our economist further limit the possibilities? A subtle and brilliant idea, one which is formalized in the above definition, is to introduce an auxiliary variable in the model. In our case this will be the *price* of the resource. We will call such auxiliary variables *latent variables*. Now by deducing, perhaps by experimentation, how much will be produced and consumed when the resource is offered at a certain price, our economist can in principle determine two maps $s: \mathbf{R}_+ \to \mathbf{R}_+$ and $d: \mathbf{R}_+ \to \mathbf{R}_+$ such that $s(p)$ equals the supply and $d(p)$ the demand when the resource has price p. This yields the behavior

$$\mathfrak{B} = \{(S, D) \in \mathbf{R}_+^2 \mid \exists\, p \in R_+ \text{ such that } S = s(p) = d(p) = D\}.$$

Note that this behavior can be viewed as being deduced from the *extended behavior*

$$\mathfrak{B}^e = \{(S,D,p) \in \mathbf{R}^3_+ \,|\, S = s(p),\, D = d(p),\, S = D\}$$

by eliminating p.

Also models with latent variables will often be described by equations

$$b_1^e(u,\,\ell) = b_2^e(u,\,\ell),$$

called *extended behavioral equations*. Here $b_1^e, b_2^e \colon \mathfrak{U} \times \mathfrak{L} \to \mathfrak{E}$. They define the extended behavior $\mathfrak{B}^e = \{(u,\ell) \in \mathfrak{U} \times \mathfrak{L} \,|\, b_1^e(u,\ell) = b_2^e(u,\ell)\}$, whence $\mathfrak{B} = P_u \mathfrak{B}^e$.

Often models are given with more structure. Relations in which $\mathfrak{U}$ is a product space occur very frequently: the purpose of a model is usually to explain the relation, the connection between variables. Of particular importance are the models in which $\mathfrak{U} = \mathfrak{U}_1 \times \mathfrak{U}_2$ and the behavior is given as the graph of a map $G \colon \mathfrak{U}_1 \to \mathfrak{U}_2$, yielding the behavioral equations

$$u_2 = G u_1$$

and $\mathfrak{B} = \{(u_1, u_2) \in \mathfrak{U} \,|\, u_2 = G u_1\}$. In these models we can view the variables u_1 as *free variables*, and the variables u_2 as *bound variables* 'caused', if you like, by u_1.

If we consider extended behavioral equations of the above type, and if we assume that the latent variables are the free variables, then we arrive at the extended behavioral equations

$$u = G^e \ell$$

with $G^e \colon \mathfrak{L} \to \mathfrak{U}$. This yields $\mathfrak{B}^e = \{(u,\ell) \in \mathfrak{U} \times \mathfrak{L} \,|\, u = G^e \ell\}$ and $\mathfrak{B} = \operatorname{im} G^e$. Hence in this case the behavior is given as the image of a map, whereas in the case of behavioral equations the behavior is specified as the kernel of a map.

Two *causes célèbres* of modelling with latent variables are probability theory and quantum mechanics. In modern probability theory one views the realization of a random variable as follows. There is an underlying space Ω and a map $r \colon \Omega \to \mathfrak{U}$. Now the gods select the basic random variable $\omega \in \Omega$, which induces the realization $u = r(\omega)$. The space Ω can be seen as a universal latent variable space. Adding measures and measurability yields the standard framework of probability theory. We will explain the situation with quantum mechanics later on.

Summarizing: A model is a subset of a universum; models are often specified by behavioral equations; models obtained from first principles will typically involve latent variables in addition to the attributes which one is trying to model.

II. Dynamical Models

5. Dynamical Systems

We will now introduce our formal definition of a dynamical system. We will view a dynamical system simply as a mathematical model in the sense explained above, but a mathematical model in which the objects of interest are functions of time. We will take the point of view that a dynamical system consists of (a family of) laws which constrain the time signals which the system can conceivably produce. The collection of all the signals compatible with these laws define what we call the *behavior* of the system. However, also here we will see that dynamical models which we write down from first principles will invariably contain, in addition to the variables which are being modelled, auxiliary variables: *latent variables.* Some latent variables may have important properties related to and capturing the memory structure of a system. This leads in particular to the concept of the *state* of a dynamical system. We will also show how systems described by difference or differential equations fit in our abstract setting.

DEFINITION. *A dynamical system* Σ *is defined as a triple*

$$\Sigma = (T, W, \mathfrak{B})$$

with

$T \subseteq \mathbf{R}$	the **time** axis;
W	an abstract set, called the **signal alphabet**;
$\mathfrak{B} \subseteq W^T$	the **behavior**.

The set T specifies the set of time instances relevant to our problem. Usually $T = \mathbf{R}, \mathbf{R}_+$ (in *continuous-time* systems), $\mathbf{Z}, \mathbf{Z}_+$ (in *discrete-time* systems) or, more generally, an interval in $\mathbf{R}$ or $\mathbf{Z}$. We view a dynamical system as an entity which is abstracted from its environment but which interacts with it. The set W specifies the way in which the attributes of the dynamical system are formalized as elements of a set. These attributes are the variables whose evolution in time we are describing. These will be a combination of observed variables and variables through which the system interacts with its environment. (If we think of the observer and the modeller as being part of the environment, then we can consider the description of this interaction with the environment as the essential feature of the attributes).

The behavior $\mathfrak{B}$ is simply a family of time trajectories taking their values in the signal alphabet. Thus elements of $\mathfrak{B}$ constitute precisely the trajectories compatible with the laws which govern the system: $\mathfrak{B}$ consists of all time signals which, according to the model, our system can conceivably generate. In most applications, the behavior $\mathfrak{B}$ will be specified by equations, often differential or difference equations, sometimes integral equations.

6. Latent Variables in Dynamical Systems

Dynamical models written down from first principles will invariably involve, in addition to the basic signals which we are trying to describe, auxiliary variables, i.e., *latent variables.* These latent variables could be introduced, if for no other reason, because they make it more convenient to write down the equations of motion or because they are essential in order to express the constitutive/conservation laws defining the system's behavior.

Latent variables will unavoidably occur whenever we model a system by *'tearing'* and *'zooming,'* in which we view a system as an interconnection of subsystems—a common and very useful way of constructing models. After interconnection, the external variables of the subsystems will become latent variables for the interconnected system.

Latent variables also play an important role in theoretical considerations. Latent variables, as state variables or free driving variables, are needed and make it possible to reduce equations of motion to expressions which are purely local in time. The formalization of systems with latent variables leads to the following definition.

DEFINITION. *A dynamical system with latent variables is a quadruple* $\Sigma_a = (T, W, A, \mathfrak{B}_a)$ *with* T, W *as before;* A *the set of* **latent variables;** $\mathfrak{B}_a \subseteq (W \times A)^T$ *the* **extended behavior.**

Define $P_w: W \times A \rightarrow W$ by $P_w(w, a) = w$. We will call Σ_a *a model with latent variables for the induced dynamical system* $\Sigma = (T, W, P_w \mathfrak{B}_a)$. Often we will refer to (and think of) $\mathfrak{B}_a$ as the *internal* behavior, and of $P_w \mathfrak{B}_a$ as the *external* behavior of the system.

We will now give two of our favorite examples illustrating these definitions of dynamical systems.

Kepler's laws. According to Kepler, the motion of planets in the solar system obey the following three laws:

K1: They move in elliptical orbits with the sun in one of the foci;

K2: The radius vector from the sun to the planet sweeps out equal areas in equal times;

K3: The square of the period of revolution is proportional to the third power of the major axis of the ellipse.

These laws define a dynamical system with $T = \mathbf{R}$, $W = \mathbf{R}^3$, and $\mathfrak{B}$ the family of all orbits satisfying K1, K2, and K3.

$$\sum_{k=1}^{n} u_i^k(t) \leq q_i(t) \qquad \forall\ i \in n \qquad (LE_1)$$

$$\sum_{j=1}^{n} a_{ij}^k\, y_j^k(t+1) \leq u_i^k(t) \qquad \forall\ k \in N,\ i \in n \qquad (LE_2)$$

$$q_i(t) \leq \sum_{k=1}^{n} y^k(t) \qquad \forall\ i \in n \qquad (LE_3)$$

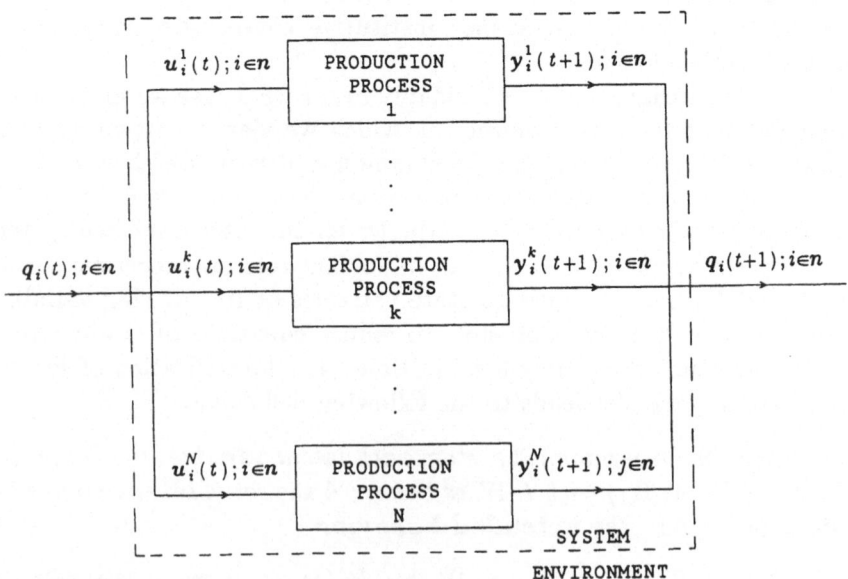

Figure 2

A Leontieff Economy. As a second example of a dynamical system let us consider a Leontieff model for an economy in which several economic goods are transformed by means of a number of production processes. We are interested in describing the evolution in time of the total utility of the goods in the economy. Assume that there are N production processes in which n economic goods are transformed into goods of the same kind, and that in order to produce one unit of good j by means of the k-th production process, we need at least a_{ij}^k units of good i. The real numbers $a_{ij}^k, k \in N, i,j \in n = \{1,2,\dots,n\}$, are called the *technology coefficients*. We assume that in each time unit one production cycle will take place.

Denote by

$$\mathbf{q}_i(t) = \text{the quantity of product } i \text{ available}$$
$$\text{at time } t;$$
$$\mathbf{u}_i^k(t) = \text{the quantity of product } i \text{ assigned}$$
$$\text{to the production process } k \text{ at time } t;$$
$$\mathbf{y}_i^k(t) = \text{the quantity of product } i \text{ acquired}$$
$$\text{from the production process } k \text{ at time } t.$$

There holds:

$$\sum_{k=1}^{n} \mathbf{u}_i^k(t) \le \mathbf{q}_i(t) \qquad\qquad \forall\, i \in \mathbf{n} \qquad\qquad \text{(LE1)}$$

$$\sum_{j=1}^{n} a_{ij}^k \mathbf{y}_j^k(t+1) \le \mathbf{u}_i^k(t) \qquad\qquad \forall\, k \in \mathbf{N},\ i \in \mathbf{n} \qquad \text{(LE2)}$$

$$\mathbf{q}_i(t) \le \sum_{k=1}^{n} \mathbf{y}_i^k(t) \qquad\qquad \forall\, i \in \mathbf{n} \qquad\qquad \text{(LE3)}$$

The underlying structure of this economy is shown in Figure 2. The difference between the right and the left hand sides of the above inequalities will be due to such things as inefficient production, imbalance of the available products, consumption, etc.

Now assume that the total utility of the goods in the economy is a function of the available amount of goods $q_1, q_2, \dots, q_n$, i.e., $\mathbf{J}: \mathbf{Z} \to \mathbf{R}_+$ is given by $\mathbf{J}(t) = \eta(\mathbf{q}_1(t), \dots, \mathbf{q}_n(t))$, with $\eta: \mathbf{R}_+^n \to \mathbf{R}_+$ a given function, the *utility*. For example, if we identify utility with resale value, then $\eta(\mathbf{q}_1, \mathbf{q}_2, \dots, \mathbf{q}_n)$ will be equal to $\sum_{k=1}^{n} p_i \mathbf{q}_i$, with p_i the per unit selling price of good i. These relations define a dynamical system with $T = \mathbf{Z}$, $W = \mathbf{R}_+$, and

$$\mathfrak{B} = \{\mathbf{J}\colon \mathbf{Z} \to \mathbf{R}_+ \mid \exists \mathbf{q}_i\colon \mathbf{Z} \to \mathbf{R}_+, \mathbf{u}_i^k\colon \mathbf{Z} \to R_+, \mathbf{y}_i^k\colon \mathbf{Z} \to \mathbf{R}_+,$$
$$i \in \mathbf{n},\ k \in \mathbf{N},\ \text{such that the inequalities (Ec)}$$
$$\text{are satisfied for all } t \in \mathbf{Z}, \text{ and } \mathbf{J} = \eta(q_1, q_2, \dots, q_n)\}\,.$$

Clearly, here it is logical to view the q_i's, u_i^k's, and y_i^k's as latent variables, and view this model as a dynamical system involving latent variables, with $T = \mathbf{Z}$, $W = \mathbf{R}_+$, $A = \mathbf{R}_+ \times \mathbf{R}_+^{n \times N}$, and

$$\mathfrak{B} = \{(\mathbf{J}, \mathbf{q}_i, \mathbf{u}_i^k, \mathbf{y}_i^k; i \in \mathbf{n},\ k \in \mathbf{N})\colon \mathbf{Z} \to W \times A \mid \text{the inequalities}$$
$$\text{(Ec) are satisfied for all } t \in \mathbf{Z}, \text{ and } \mathbf{J} = \eta(\mathbf{q}_1, \mathbf{q}_2, \dots, \mathbf{q}_n)\}.$$

7. Dynamical Behavioral Equations

Just as in the case of general mathematical models, the (extended) behavior is usually specified by behavioral equations. Since these equations will involve time, we may want to refer to them as *dynamical behavioral equations.* In dynamical systems these equations are often differential or difference equations.

Dynamical behavioral equations described by difference equations are given by a map $f\colon W^{L+1} \to \mathbf{R}^g (g \in \mathbf{N}$ equals the number of equations specifying the behavior). The resulting difference equation is

$$f(\mathbf{w}(t + L), \mathbf{w}(t + L - 1), \dots, \mathbf{w}(t)) = 0.$$

This equation can be written more compactly by using the (*backwards*) *shift* σ defined by $(\sigma \mathbf{w})(t) = \mathbf{w}(t + 1)$. This yields

$$f \circ (\sigma^L \mathbf{w}, \sigma^{L-1} \mathbf{w}, \dots, \mathbf{w}) = 0. \qquad (*)$$

With latent variables, we obtain

$$f^e \circ (\sigma^L \mathbf{w}, \sigma^{L-1} \mathbf{w}, \dots, \mathbf{w}, \sigma^{L_a} \mathbf{a}, \sigma^{L_a - 1} \mathbf{a}, \dots, \mathbf{a}) = 0.$$

The formal definition of the system induced by these equation is quite obvious. Take $T = \mathbf{Z}$ or $\mathbf{Z}_+$, and

$$\mathfrak{B} = \{\mathbf{w}\colon T \to W \mid \mathbf{w} \text{ satisfies } (*)\},$$

with obvious generalizations to latent variables.

In the differential equation case, the analogous behavioral equation takes the form

$$f \circ \left(\frac{d^L \mathbf{w}}{dt^L}, \frac{d^{L-1}\mathbf{w}}{dt^{L-1}}, \ldots, \mathbf{w} \right) = \mathbf{0}.$$

The precise definition of the behavior and the sense in which this differential equation needs to be satisfied is a mathematical issue which we will not pursue here. In fact, it is often convenient and necessary to interpret the $\mathbf{w}$'s and the equations in the sense of distributions.

8. Linearity and Time-Invariance

One of the advantages of making definitions at the level of generality as done here is that standard mathematical structures become immediately applicable to dynamical systems.

We will call the dynamical system $\Sigma = (T, W, \mathfrak{B})$ *linear* if W is a vector space and $\mathfrak{B}$ is a linear subspace of W^T (viewed as a vector space in the natural way by pointwise addition and scalar multiplication).

We will call the dynamical system $\Sigma = (T, W, \mathfrak{B})$ *time-invariant* if T is an additive semigroup in $\mathbf{R}$ (i.e., $t_1, t_2 \in T \Rightarrow t_1 + t_2 \in T$) and $\sigma^t \mathfrak{B} \subseteq \mathfrak{B}$ for all $t \in T$; σ^t denotes the *backwards* or *left t-shift*: $(\sigma^t \mathbf{f})(t') = \mathbf{f}(t' + t)$. In the sequel we will only consider time-invariant systems.

We end this section with another example.

Quantum mechanics warns us not to speak lightly about the position of a particle as a physical reality, but instead to consider the 'probability' of finding a particle in a certain region of space $\mathbf{R}^3$. Thus we will obtain a dynamical system with time axis $T = \mathbf{R}$ and signal alphabet $P = \{p : \mathbf{R}^3 \to \mathbf{R} \mid p \geq 0 \text{ and } \int_{\mathbf{R}^3} p(z)\,dz = 1\}$—the collection of all probability measures (which for simplicity we have taken to be absolutely continuous w.r.t. Lebesgue measure) on $\mathbf{R}^3$. In order to specify the behavior, it has proven to be convenient to introduce the *wave function* $\psi : \mathbf{R}^3 \to \mathbf{C}$ as a latent variable. Thus define the space of latent variables $\Psi = \mathfrak{L}_2(\mathbf{R}^3; \mathbf{C})$. The internal behavior $\mathfrak{B}_a \subseteq (P \times \Psi)^{\mathbf{R}}$ is defined by two dynamical behavioral equations. The first determines p as a function of ψ, while the second, the *Schrödinger equation*, tells us how ψ evolves in time. Let $\boldsymbol{\psi} : \mathbf{R}^3 \times \mathbf{R} \to \mathbf{C}$ be the time trajectory of the wave function. Hence $\boldsymbol{\psi}(z_1, z_2, z_3; t)$ denotes the value of the wave function at the position $(z_1, z_2, z_3) \in \mathbf{R}^3$ at time $t \in \mathbf{R}$. Similarly, let $\mathbf{p} : \mathbf{R}^3 \times \mathbf{R} \to \mathbf{R}_+$ denote the time trajectory of the probability density function. The wave function generates the probability density by

$$\mathbf{p}(z_1, z_2, z_3; t) = \frac{|\psi(z_1, z_2, z_3; t)|^2}{\int_{\mathbf{R}^3} |\psi(z_1, z_2, z_3; t)|^2 \, dz_1 \, dz_2 \, dz_3} \qquad \text{(QM1)}$$

The evolution of the wave function is governed by Schrodinger's equation:

$$-i\frac{h}{2\pi}\frac{\partial\psi}{\partial t} = H(\psi), \qquad \text{(QM2)}$$

where the *Hamiltonian H* is a linear, in general unbounded, operator on $\mathfrak{L}_2(\mathbf{R}^3; \mathbf{C})$ and h is Planck's constant. The Hamiltonian is specified by the potential and the geometry and should be considered as fixed for a given system. This yields the extended behavior

$$\mathfrak{B}^e = \{(\mathbf{p}, \psi) : \mathbf{R} \to P \times \Psi | \text{ (QM) is satisfied}\},$$

which we view as a convenient way of specifying the external behavior

$$\mathfrak{B} = \{\mathbf{p} : \mathbf{R} \to P \,|\, \exists\, \psi : \mathbf{R} \to \Psi \text{ such that (QM) is satisfied}\}.$$

Clearly this system $(\mathbf{R}, W, \mathfrak{B})$ is time-invariant. If we restrict our attention to the wave function alone, i.e., if we consider the dynamical system $(\mathbf{R}, \Psi, P_\psi \mathfrak{B}^e)$, then we obtain a linear system.

The viewpoint taken here, in which ψ is a latent variable aimed at modelling p, is a very logical one indeed. The truly surprising fact however is that the (very *nonlinear*) behavior $\mathfrak{B}$ can be represented by means of a *linear* flow (QM2), the Schrödinger equation, together with the *memoryless* map, the static behavioral equation (QM1). Note, however, that the point of view that ψ is introduced in order to model p, however logical, does not do justice to the historical development in which ψ had been studied long before the probability interpretation of $|\psi|^2$ was suggested. Note also that our approach discusses probability in a purely deterministic tone—stochastic generalizations are another story altogether.

Finally, it remains a very valid question as to whether there are other convenient ways of expressing the behavior $\mathfrak{B}$ which, instead of having to consider the difficult-to-explain wave function, are based on more palatable latent variables.

Summarizing: A dynamical system is simply a family of time trajectories. The system is time-invariant if this family is shift-invariant and linear if this family forms a linear subspace. Dynamical systems are often described by behavioral equations which are difference or differential equations. Dynamical models obtained from first principles usually also contain latent variables.

III. Linear Models

9. Introduction

Many dynamical systems encountered in practice, in particular in control, signal processing, econometrics, electrical circuit design, etc. can be effectively modelled by a finite set of linear differential or difference equations. These models may or may not contain latent variables, they may or may not display their state, and they may or may not exhibit their input/output structure. A consistent theory should be able to start from any set of dynamical equations and extract further structure, if needed, from there.

We will introduce these systems from an abstract point of view. For simplicity of notation, we will only consider discrete-time systems. Most of the results generalize easily to continuous-time systems (with the shift σ interpreted as differentiation $\frac{d}{dt}$). Also we will assume that the time axis $T = \mathbf{Z}$, and that we are considering a system which can be described by a finite set of real valued attributes: $W = \mathbf{R}^q$.

Three natural properties of a dynamical system $\Sigma = (\mathbf{Z}, \mathbf{R}^q, \mathfrak{B})$ entirely characterize the class of dynamical system under consideration. They are linearity, time-invariance, and completeness. We will call Σ *time-invariant* if $\sigma\mathfrak{B} = \mathfrak{B}$, *linear* if $\mathfrak{B}$ is a linear subspace of $(\mathbf{R}^q)^{\mathbf{Z}}$, and *complete* if $\{\mathbf{w} \in \mathfrak{B}\} \Leftrightarrow \{\mathbf{w}|_{[t_0,t_1]} \in \mathfrak{B}|_{[t_0,t_1]}$ for all $t_0, t_1 \in \mathbf{Z}\}$. Time-invariance says that the laws of a dynamical system do not change in time. Linearity requires the superposition principle to hold. Completeness states that the laws governing the behavior of the system do not involve the behavior of the signals near $-\infty$ and/or $+\infty$.

These types of systems admit a convenient mathematical characterization within the class of dynamical systems.

PROPOSITION. $\Sigma = (\mathbf{Z}, \mathbf{R}^q, \mathfrak{B})$ *defines a linear, time-invariant, complete system iff* $\mathfrak{B}$ *is a linear, shift-invariant, closed subspace of* $(\mathbf{R}^q)^{\mathbf{Z}}$, *equipped with the topology of pointwise convergence.*

10. Parametrization

In this section we will discuss parametrizations of linear, time-invariant, complete systems. By a parametrization, we understand the following. Let S be a set consisting of *'abstract'* objects and $\mathcal{P}$ a set consisting of *'concrete'* objects. Let π be a surjective map $\pi: \mathcal{P} \to S$. The pair $(\mathcal{P}, \pi)$ is called a *parametrization* of S. If $\pi(p) = s$, then we will call $p \in \mathcal{P}$ a *parametrization* of $s \in S$. Note that π need not be injective.

Thus π induces an equivalence relation, ker π, defined by

$$\{p_1 \sim p_2\}: \Leftrightarrow \{\pi(p_1) = \pi(p_2)\}.$$

A *canonical form* is a subset $\mathcal{P}' \subseteq \mathcal{P}$ such that $\mathcal{P}'$ contains at least one element of each equivalence class. In other words, a canonical form requires that $\pi|_{\mathcal{P}'} : \mathcal{P}' \to \mathcal{S}$ remains surjective. If $\mathcal{P}'$ contains exactly one element of each equivalence class, we will speak of a *trim canonical form*. Hence $\mathcal{P}'$ is a trim canonical form iff $\pi|_{\mathcal{P}'} : \mathcal{P}' \to \mathcal{S}$ is a bijection.

The natural equivalence relation ker π on $\mathcal{P}$ leads to invariants. Hence an *invariant* for a parametrized class of systems is a map $i : \mathcal{P} \to I$ such that $\{p_1 \sim p_2\} \Rightarrow \{i(p_1) = i(p_2)\}$. If $\{p_1 \sim p_2\} \Leftrightarrow \{i(p_1) = i(p_2)\}$, then we will call i a *complete invariant*.

11. AR and ARMA Models

We start this section with a few words about notation. We will (as above) employ the standard notation: $\mathbf{R}$ for the real line, $\mathbf{R}_+$ for $[0, \infty]$, $\mathbf{Z}$ for the integers, $\mathbf{Z}_+$ for $\{0, 1, 2, \dots\}$, $\mathbf{R}^n$ for real n-dimensional vector space, $\mathbf{R}^{n_1 \times n_2}$ for the real $n_1 \times n_2$ matrices, $\mathbf{C}$ for the complex plane, etc. Further, W^T denotes the collection of all maps from T into W. Hence $(\mathbf{R}^q)^{\mathbf{Z}}$ denotes the collection of all q-dimensional real time-series defined on the integers. Also, $\mathbf{R}[s]$ denotes the real polynomials in the indeterminate s and $\mathbf{R}^{n_1 \times n_2}[s]$ denotes the real $n_1 \times n_2$ matrix polynomials in the indeterminate s. Similarly $\mathbf{R}[s, s^{-1}]$ denotes the real 'polynomials' containing both positive and negative powers of s; $\mathbf{R}^{n_1 \times n_2}[s, s^{-1}]$ is analogously defined. $\mathbf{R}(s)$ denotes the real rational functions in the indeterminate s. Observe that $\mathbf{R}(s)$ is the fraction field induced by both the rings $\mathbf{R}[s]$ and $\mathbf{R}[s, s^{-1}]$. So the fact that we will be considering polynomials with both positive and negative powers will have no consequence for the rational functions and, hence, for transfer functions. Finally, ker denotes the kernel (= the nullspace) of a linear map, while im denotes the image (= the range) of a map.

Let us take a look at two classes of dynamical models. These are described by systems of difference (or differential) equations. Such systems of difference equations can be written compactly using polynomial operators in the shift. We will find it convenient to use both the left shift (σ) and the right shift (σ^{-1}). Thus if $P(s, s^{-1}) = P_\ell s^\ell + P_{\ell-1} s^{\ell-1} + \dots + P_{\ell'+1} s^{\ell'+1} + P_{\ell'} s^{\ell'}$ is a polynomial matrix, an element of $\mathbf{R}^{g \times q}[s, s^{-1}]$, then $P(\sigma, \sigma^{-1})$ denotes the map from $(\mathbf{R}^q)^{\mathbf{Z}}$ into $(\mathbf{R}^g)^{\mathbf{Z}}$ defined by $(P(\sigma, \sigma^{-1})\mathbf{w})(t) = P_\ell \mathbf{w}(t + \ell) + P_{\ell-1}\mathbf{w}(t + \ell - 1) + \dots + P_{\ell'+1}\mathbf{w}(t + \ell + 1) + P_{\ell'}\mathbf{w}(t + \ell')$.

ARMA-models. Let

$$R(s,s^{-1}) \in \mathbf{R}^{g\times q}[s,s^{-1}], \qquad M(s,s^{-1}) \in \mathbf{R}^{g\times d}[s,s^{-1}],$$

and consider the equation

$$R(\sigma,\sigma^{-1})\mathbf{w} = M(\sigma,\sigma^{-1})\mathbf{a}. \qquad \textbf{(ARMA)}$$

This equation can be viewed as a law linking the q-dimensional signal time-series $\mathbf{w}: \mathbf{Z} \to \mathbf{R}^q$ to the d-dimensional latent variable time-series $\mathbf{a}: \mathbf{Z} \to \mathbf{R}^d$.

AR-models. Let $R(s,s^{-1}) \in \mathbf{R}^{g\times q}[s,s^{-1}]$ and consider the equation

$$R(\sigma,\sigma^{-1})\mathbf{w} = 0 \qquad \textbf{(AR)}$$

viewed as a law restricting the signal time-series $\mathbf{w}: \mathbf{Z} \to \mathbf{R}^q$.

Let us consider the models (ARMA) and (AR). The first of these defines a model with latent variables. Its full behavior consists of the set of time series $(\mathbf{w},\mathbf{a})$ satisfying (ARMA). In mathematical notation, this set is of course nothing other than the kernel of the map $[R(\sigma,\sigma^{-1}) \mid -M(\sigma,\sigma^{-1})]$.

Formally, (ARMA) defines thus the dynamical system with latent variables $\Sigma_a = (\mathbf{Z},\mathbf{R}^q,\mathbf{R}^d,\mathfrak{B}_a)$, with $\mathfrak{B}_a = \ker [R(\sigma,\sigma^{-1}) \mid -M(\sigma,\sigma^{-1})]$. Similarly, (AR) defines the (intrinsic) dynamical system $\Sigma = (\mathbf{Z},\mathbf{R}^q,\mathfrak{B})$ with $\mathfrak{B} = \ker R(\sigma,\sigma^{-1})$.

Three questions arise:

1. Are these two classes of systems really different?

2. Is it possible to give some natural system-theoretic properties of a dynamical system $\Sigma = (\mathbf{Z},\mathbf{R},\mathfrak{B})$ which will allow it to be represented by means of an (AR) model?

3. When do two polynomial matrices R_1 and R_2 define (AR) systems with the same behavior?

The answer to the first question is 'No' and to the second question is 'Yes.' The sense in which an (ARMA) system may be represented as an (AR) system is easy to state (and easy to prove—although we will dispense with proofs). Indeed:

PROPOSITION. *Let $\Sigma_a = (\mathbf{Z},\mathbf{R}^q,\mathbf{R}^d,\mathfrak{B}_a)$ be an (ARMA) dynamical system with latent variables. Then there exists an (AR) system $\Sigma = (\mathbf{Z},\mathbf{R}^q,\mathfrak{B})$ such that $\mathfrak{B}$ is the external behavior induced by $\mathfrak{B}_a$.*

What does this proposition say? In mathematical terms, it states that for any $R(s,s^{-1}) \in \mathbf{R}^{g \times q}[s,s^{-1}]$ and $M(s,s^{-1}) \in \mathbf{R}^{g \times d}[s,s^{-1}]$, there exists a g' and an $R'(s,s^{-1}) \in \mathbf{R}^{g' \times q}[s,s^{-1}]$ such that

$$(R(\sigma,\sigma^{-1}))^{-1} \operatorname{im} (\sigma,\sigma^{-1}) = \ker R'(\sigma,\sigma^{-1}).$$

(Here $(\cdot)^{-1}$ denotes the set-theoretic inverse). More down to earth, this says that the set of time-series $\mathbf{w}: \mathbf{Z} \to \mathbf{R}^q$ such that $R(\sigma,\sigma^{-1})\mathbf{w} = M(\sigma,\sigma^{-1})\mathbf{a}$ for some $\mathbf{a}: \mathbf{Z} \to \mathbf{R}^d$ will be exactly those $\mathbf{w}: \mathbf{Z} \to \mathbf{R}^q$ such that $R'(\sigma,\sigma^{-1})\mathbf{w} = \mathbf{0}$.

This fact, implying that for the class of systems under consideration the latent variables can always be eliminated from the model equations, is of considerable interest in modelling. Indeed, many modelling procedures are based on *tearing:* a system is viewed as an interconnection of a number of subsystems, and modelling is done by *zooming-in* on the individual subsystems. This will yield a model which, of course, will necessarily contain the interconnecting variables as latent variables. The fact that these interconnecting variables may be eliminated implies that the complexity of the final system will, as far as the number of variables is concerned, be influenced only by the number of variables being modelled, and not by the number of subsystems generated by the tearing process.

It turns out that (AR) models are precisely those introduced from an abstract point of view in Section 9. Indeed:

PROPOSITION. $\Sigma = (\mathbf{Z}, \mathbf{R}^q, \mathfrak{B})$ *is linear, time-invariant, and complete if and only if there exists a g and a polynomial matrix $R(s,s^{-1}) \in \mathbf{R}^{g \times q}[s,s^{-1}]$ such that $\mathfrak{B} = \ker R(\sigma,\sigma^{-1})$.*

Let us now turn to the third question. Different (AR) models may have the same behavior. It is possible to express this condition in terms of the (AR) matrices. We will call two polynomial matrices $R_1(s,s^{-1})$, $R_2(s,s^{-1}) \in \mathbf{R}^{g \times q}[s,s^{-1}]$ *unimodularly equivalent* if there exists a polynomial matrix $U(s,s^{-1}) \in \mathbf{R}^{g \times g}[s,s^{-1}]$ which is *unimodular* (that is, $\det U(s,s^{-1}) = \alpha s^d$ for some $\alpha \neq 0$ and $d \in \mathbf{Z}_+$—equivalently, U is invertible as a *polynomial matrix*) such that $R_2 = UR_1$. If $R_1(s,s^{-1}) \in \mathbf{R}^{g_1 \times q}[s,s^{-1}]$ and $R_2(s,s^{-1}) \in \mathbf{R}^{g_2 \times q}[s,s^{-1}]$, with $g_1 \neq g_2$, then we will call them *unimodularly equivalent* if, after adding zero rows to R_1 and/or R_2 so as to make them have the same number of rows, they are unimodularly equivalent in the above sense. It is easy to prove that a polynomial matrix is unimodularly equivalent to one which is of *full row rank* (that is, $R(s,s^{-1}) \in \mathbf{R}^{g \times q}[s,s^{-1}]$ has rank g over $\mathbf{R}(s)$).

PROPOSITION. *Two (AR) systems with defining polynomial matrices* $R_1(s, s^{-1})$ *and* $R_2(s, s^{-1})$ *have the same behavior if and only if* R_1 *and* R_2 *are unimodularly equivalent.*

The above propositions imply that (AR) models parametrize the linear, time-invariant, complete systems. The resulting space of parameters is $\mathbf{R}^{\cdot \times q}[s, s^{-1}]$—the polynomial matrices with q columns. It is possible to limit the set of parameters further. Canonical forms are $\mathbf{R}^{q \times q}[s, s^{-1}]$, or even $\mathbf{R}_f^{\cdot \times q}[s]$, the full row rank polynomial matrices with q columns. The construction of a trim canonical form and of a complete set of invariants is a delicate and interesting issue. Details are given in the references.

Summarizing: Linear, time-invariant, complete dynamical systems have behaviors which are kernels of polynomial operators in the shift. Such systems are, hence, parametrized by polynomial matrices. We call the resulting dynamical behavioral equations (AR) models. Introducing latent variables in this structure leads to (ARMA) models. The latent variables can always be eliminated, since (ARMA) models define the same class of systems as (AR) models.

IV. The Structure of the Behavior

12. Well-Posedness

We have all been raised with the principle, ascribed to Hadamard, that *'well-posed'* mathematical models should have unique solutions for given initial and boundary conditions, and that the solutions should depend continuously on these initial conditions and on the parameters of the model. The last requirement has long been abandoned with the advent of bifurcation theory, chaos, and the like.

The uniqueness condition is also strange. At best, the principle is a tautology in the sense that it *defines* the initial and boundary conditions as those needed to yield this uniqueness.

From a system-theoretic point of view, the interesting part of a dynamical system is its interaction with its environment. The environment will interact with the system in a manner which is in principle free and unpredictable. No matter how many 'internal' initial conditions are specified, no uniqueness will result. It is in this way that our concept of a dynamical system differs radically from the classical one.

In this classical concept, a dynamical system is described in terms of the evolution of its state. It is assumed that the state evolves in an autonomous way. By this, we mean that its path depends only on its

initial value and on the laws of motion. In situations other than some
very well defined mechanical or electrical systems, the theory leaves us
guessing as to how the state variables should be chosen. Further, no
external influences are formally incorporated into this framework: the
state evolves purely on the basis of internal driving forces.

By assuming that the state evolves in this deterministic fashion we
postulate, in effect, that the system is isolated from its environment.
But there is no such thing as an isolated system! What this assumption
actually means is that we postulate that we know, or think we know,
how the environment will act on the system, what the boundary conditions are, how external influences are generated; thus, in modelling
a specific, concrete, dynamical system in the language of classical dynamics, *we find ourselves in the absurd situation of having to model the
environment as well!*

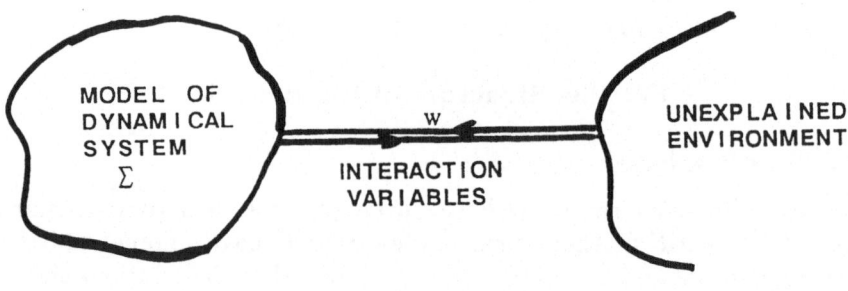

Figure 3

Our formalization of a dynamical system stems from the mental
picture shown in Figure 3. We view a dynamical system as an object
which is imbedded in its environment, is abstracted from it, but which
may (and will) interact with it. The system has certain attributes
whose evolution in time we desire to describe. The dynamical laws
specifying this time evolution tell us that certain trajectories can occur
and that others cannot. This yields what we have called the *behavior*
of the dynamical system. This point of view takes the model equations,
any set of dynamical relations, as basic and proceeds from there. That
is what the modeller gives us, that is what a mathematical theory of

dynamics should start with. If preconditioning of the model is necessary (for example, in order to display the evolution of the state or the input/output structure), then a theory should make clear how and why this should be done.

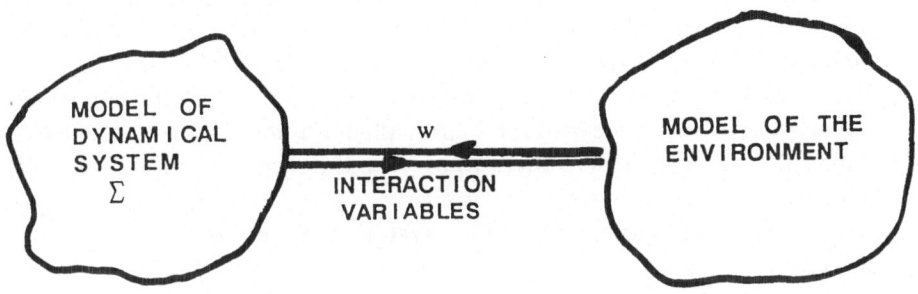

Figure 4

Note that it is only by viewing the environment, too, as a specific dynamical system that we are able to obtain an autonomous classical dynamical system. It is in this sense that classical dynamics forces us to model also the environment (see Figure 4).

The solution set of a dynamical system will usually have the following structure: there will be free initial conditions and a free input signal. A detailed analysis of this situation is given in [2]. Here we will content ourselves just with explaining the situation for linear systems.

13. Input/State/Output Models

Let $A \in \mathbf{R}^{n \times n}, B \in \mathbf{R}^{n \times m}, C \in \mathbf{R}^{p \times n}$, and $D \in \mathbf{R}^{p \times m}$ and consider the ubiquitious systems of equations

$$\sigma \mathbf{x} = A\mathbf{x} + B\mathbf{u}$$
$$\mathbf{w} = \begin{pmatrix} \mathbf{u} \\ \mathbf{y} \end{pmatrix}$$
$$\mathbf{y} = C\mathbf{x} + D\mathbf{u} \qquad \text{(i/s/o)}$$

linking the state time-series $\mathbf{x}\colon \mathbf{Z} \to \mathbf{R}^n$, the output time-series $\mathbf{y}\colon \mathbf{Z} \to \mathbf{R}^p$, and the input time-series $\mathbf{u}\colon \mathbf{Z} \to \mathbf{R}^m$.

The above system of equations has a very special structure. When viewed as an equation in $\mathbf{x}$ and $\mathbf{w}$, it is first order in $\mathbf{x}$ and zero-th order in $\mathbf{w}$. This implies that $\mathbf{x}$ is a *state* variable, meaning that any trajectory terminating at a state $\mathbf{x}(0)$ can be concatenated with any trajectory emanating from this same state $\mathbf{x}(0)$ and yield a trajectory satisfying the dynamical equations. For a formalization of the *axiom of state*, we refer the reader to the references. Viewed as an equation in $\mathbf{u}$, (i/s/o) has some special properties, implying that $\mathbf{u}$ is completely unconstrained by these equations. More specifically, for any $\mathbf{u} \colon \mathbf{Z} \to \mathbf{R}^m$, there will exist $\mathbf{x} \colon \mathbf{Z} \to \mathbf{R}^n$, and $\mathbf{y} \colon \mathbf{Z} \to \mathbf{R}^p$ such that the equations are satisfied. Such free variables are called *inputs*. Once $\mathbf{u} \colon \mathbf{Z} \to \mathbf{R}^m$ and $\mathbf{x}(0) \in \mathbf{R}^n$ are chosen, the *output* $\mathbf{y}$ and, hence, $\mathbf{w}$ will be uniquely defined.

It turns out that every (AR) system may be written this way. Indeed:

THEOREM. *Let* $\Sigma = (\mathbf{Z}, \mathbf{R}^q, \mathfrak{B})$ *be a linear, time-invariant, complete system. Then there exists a* $q \times q$ *permutation matrix* π, *integers* m, p *with* $m + p = q$, *an integer* n, *and matrices* A, B, C, D *such that* $\pi \mathfrak{B}$ *equals the external behavior of the dynamical system with latent variables described by* (i/s/o).

Observe that in most models it will not be clear what the relation is between the signal $\mathbf{w}$ and the state $\mathbf{x}$, nor will it be clear what components of $\mathbf{w}$ constitute the input $\mathbf{u}$. Constructing the state is the subject of realization theory (see the references).

The important conclusion is that even when $\mathbf{x}(0)$ is specified, there still will be free input variables which are not, and need not be, explained by the dynamical models. There is certainly no need and no justification at all for specifying them as random processes. In our opinion, the ideology that a model must yield a unique solution is one of the most serious pitfalls of the generally accepted modelling philosophy. It leads to the need, often unjustifiable, of having to explain external forces as, for example, in assuming that in economic models exogenous variables are random variables.

Note finally that the classical models in physics contain inputs. This is clear in Newton's second law

$$M \frac{d^2 \vec{\mathsf{q}}}{dt^2} = \vec{\mathbf{F}}$$

and in Maxwell's equations.

Maxwell's equations in free space read:

$$\nabla \cdot \vec{E} = \frac{\rho}{\epsilon_0} \qquad \text{(EM1)}$$

$$\nabla \times \vec{E} = -\frac{\partial \vec{B}}{\partial t} \qquad \text{(EM2)}$$

$$\nabla \cdot \vec{B} = 0 \qquad \text{(EM3)}$$

$$c^2 \nabla \times \vec{B} = \frac{\vec{j}}{\epsilon_0} + \frac{\partial \vec{E}}{\partial t} \qquad \text{(EM4)}$$

Here $\nabla\cdot$ denotes the divergence, $\nabla\times$ the curl, ϵ_0 the dielectric constant of free space, and c the speed of light.

Maxwell's equations consist of two static (EM1 and EM3) and two dynamic (EM2 and EM4) behavioral equations. These imply the law of conservation of charge:

$$\nabla \cdot \vec{j} = -\frac{\partial \rho}{\partial t} . \qquad \text{(EM5)}$$

It is easy to see that Maxwell's equations should be viewed as a dynamical system with $\vec{j}$ as an (unexplained) input variable.

It's worth making a slightly philosophical comment. Maxwell's equations (EM1, EM2, EM4) by themselves define an i/s/o system with $\vec{j}$ and ρ jointly as input variables. It is reasonable and logical to consider these equations as a physical reality, regardless of whether or not (EM3) and, hence, (EM5) is (assumed to be) satisfied. If we do so, then both $\vec{j}$ and ρ are viewed as being imposed by of the environment. Equation (EM5) may be regarded on its own as an i/s/o variable with $\vec{j}$ as input variable and ρ as output variable, which partly explains this environment. The interconnection of both systems yields a dynamical system with only input. In his well known lecture notes, Feynman argues that it makes no physical sense to claim that only a subset of these equations are true without incorporating the others. From a system-theoretic vantage point, we feel little sympathy for this point of view. *Is it really necessary to consider all the laws of physics all at once in every single investigation? Is it impossible to declare part of them as being true while disregarding the others?* The position is obviously untenable. The system-theoretic, reductionist point of view, in which it is allowed to leave the environment totally unexplained (even though more scrutiny will undoubtedly lead to the discovery of additional relations), is much more reasonable. Any set of dynamical relations defines a dynamical system, a reality in its own right. It will usually contain unexplained inputs. Further analysis may lead to the discovery of more relations, leaving fewer unexplained inputs. This further analysis can, but need not be done. Otherwise we will always end up having to model the whole universe.

Summarizing: In dynamical models there will be free initial conditions and free variables imposed by the environment. There is no way and no need for the dynamical model to explain these.

V. Modelling From Data

14. General Ideas

We will now address the question of identification from measurements on a phenomenon that we want to obtain a model for. Even though we are especially interested in linear dynamical systems, we will explain our ideas in full generality.

Influenced by the thinking and descriptive methodology of physics, our thoughts about the modelling process usually proceed as follows. We start with a system consisting of a family of interconnected subsystems. Then, based on physical principles and laws, we set up models for these subsystems and for their interconnections. Typically, the laws of the subsystems will involve the constitutive equations of the elements of the the system, while those of the interconnections will often only involve field equations and conservation laws. Perhaps there may also be a number of numerical element values in the constitutive laws which we are not sure about. These may be left as parameters or will be determined by means of measurements. Putting this all together leads to the standard modelling procedures. The traditional laws of physics play the key role in this methodology, while observations play an important but ancillary role.

However, in other areas such as econometrics, signal processing, (adaptive) control, etc., there is an increasing emphasis on modelling procedures in which the model is obtained directly and almost solely from the observed data, and in which the model will be automatically updated and perhaps restructered as more and newer measurements about the process become available. *The idea is to let the data speak,* as they say. Models and laws obtained in this way should in the first instance be considered as descriptive (they organize the observations and can be viewed as data compression) but not as interpretative models. Nevertheless such models may have very good predictive value.

The field of applied mathematics which is directly concerned with modelling from data is statistics. In many ways statistics, with its beautiful interplay of concepts, mathematics, algorithms, and applications, can be viewed as a paradigm for applied mathematics. For good historical reasons and some impressive success stories, statistics has transformed the problem of modelling into a problem of stochastics, and has chosen the question of reconstructing a probability law from a finite sample as the basic question in modelling from data. This vantage point is very restrictive, and in our opinion plays much too central a role. This starting point has also dominated modelling procedures in fields like econometrics, where even the stochastic nature of the phenomena is questionable and certainly not compelling. As such, the field of modelling from data has been totally dominated by a philosophy which equates uncertainty with stochasticity, which identifies unknown with random, and which reduces data to samples.

In this section we will explain a 'deterministic' approach to the question of modelling from data, with particular attention paid to the problem of finding dynamical models from observed time-series. As is customary in control and system theory, we will refer to this as the *identification problem.*

Assume that we are trying to model a *phenomenon* described as in Section 2 by means of some attributes which we have quantified as elements of a set $\mathfrak{U}$. In Section 2 we have explained how mathematical models fit into this setting. In order to study the problem of modelling from data, we have to incorporate *measurements* into this framework. We view measurements as a subset of a space $\mathfrak{D}$. We assume, of course, that these measurements are related to the phenomenon which is being modelled. Now, on the one hand, identical measurements may result from different realizations of the phenomenon; on the other hand, the measurements may be influenced by factors other than the phenomenon itself, and so identical realizations of the phenomenon may also lead to different measurements. We can incorporate this fact by assuming that the measurements are influenced by latent variables. Often, of course, the latent variables which influence the measurements will be distinct from those incorporated in the model itself.

A model, together with a measurement process, is thus formalized as $(\mathfrak{U}, \mathfrak{D}, \mathfrak{L}, \mathfrak{B}^e, d)$ with $\mathfrak{U}$ the space of *attributes of the phenomenon,* $\mathfrak{L}$ the space of *latent variables,* $\mathfrak{B}^e \subseteq \mathfrak{U} \times \mathfrak{L}$ the *model,* and $d: \mathfrak{U} \times \mathfrak{L} \to \mathfrak{D}$ the *measurement map.*

We will now consider the problem of constructing a model on the basis of (observed or postulated) measurements. We are trying to model a phenomenon with attributes in a set $\mathfrak{U}$ on the basis of data in a set $\mathfrak{D}$. We postulate a model class M. Each element of M is thus a quintuple $(\mathfrak{U}, \mathfrak{D}, \mathfrak{L}, \mathfrak{B}^e, d)$. The specification of the model class will incorporate *a priori* conditions on the model. These may be conditions of a pragmatic nature (as linearity) or requirements stemming from physical laws (symmetries, conservation laws). The specification of M could also incorporate the nature of the measurement map d if it happens to be known. The problem of obtaining a model on the basis of observations may now be formalized as follows. Let $\mathfrak{D}'$ be a given subset of $\mathfrak{D}$; $\mathfrak{D}'$ represents the observed realizations of the measurement process. The problem is to deduce from $\mathfrak{D}'$ a model $(\mathfrak{U}, \mathfrak{D}, \mathfrak{L}, \mathfrak{B}^e, d) \in M$.

Let D denote the class of possible measurement sets which will be considered. Often D incorporates certain finiteness or boundedness restrictions which we have to postulate on measurements in order for algorithms to be well-defined. The question of modelling on the basis of data comes down to constructing a map $i: D \to M$. We can think of i as an *identification procedure*. An important consideration in deducing a model is the concept of *falsification*. The model $(\mathfrak{U}, \mathfrak{D}, \mathfrak{L}, \mathfrak{B}^e, d)$ and a measurement set $\mathfrak{D}' \subseteq \mathfrak{D}$ define an *unfalsified model* if $\mathfrak{D}' \subseteq d(\mathfrak{B}^e)$.

In order to illustrate these abstract ideas, let us apply it to the setting of linear, time-invariant, complete systems and restrict attention to the case in which we are trying to model the observations directly. This corresponds to assuming $\mathfrak{D} = \mathfrak{U}$ and assuming that the measurement map is given by $d: (u, \ell) \mapsto u$. Let us also assume that we have observed a finite set of time-series $\tilde{\mathbf{w}}_i: \mathbf{Z} \to \mathbf{R}^q$, $i \in \{1, 2, \ldots, k\}$. Usually, in fact, $k = 1$: we will be trying to identify the system on the basis of only one response function. Hence the model M class corresponds to the (ARMA) models and consists of all finite subsets of $(\mathbf{R}^q)^{\mathbf{Z}}$. Thus each model in M tries to explain the k observations $\{\tilde{\mathbf{w}}_1, \tilde{\mathbf{w}}_2, \ldots, \tilde{\mathbf{w}}_k\}$ by means of two polynomial matrices $R(s, s^{-1}) \in \mathbf{R}^{g \times q}[s, s^{-1}]$ and $M(s, s^{-1}) \in \mathbf{R}^{g \times d}[s, s^{-1}]$ in the sense that it proposes the equations

$$R(\sigma, \sigma^{-1})\mathbf{w} = M(\sigma, \sigma^{-1})\mathbf{a} \qquad \text{(ARMA)}$$

as a reasonable way of interpreting the $\tilde{\mathbf{w}}_i$'s.

The question thus becomes:

Which (ARMA) model should we choose as the most reasonable explanation of the observed $\tilde{\mathbf{w}}_i$'s?

15. The Most Powerful Unfalsified Model

The guiding principles for constructing models are: *low complexity, low misfit,* and *low latency.* The *complexity* is a measure which tells us how much a model can explain. The more complex a model is, the more it explains; the less predictive possibilities it has, the less desirable it is. The *misfit* measures to what degree the data corroborates the measurements. The *latency* measures the importance of the unexplained, or at least unobserved, latent variables. Models which corroborate the measurements well, without having to rely on latent variables with large values, inspire *prima facie* more confidence than models which require these latent variables to take on large values.

In the present section we will consider the case in which the model is required to be *unfalsified* by the data (no misfit) and in which this exact fit is achieved without relying on latent variables (no latency). We will look for the least complex (unfalsified) model, which we will interpret as the *most powerful one.*

There are two possible situations: either the most powerful unfalsified model (MPUM) exists, or it does not. We will see that in the case of linear, time-invariant, complete systems, the MPUM indeed exists. That is in itself a surprising and useful result—we will explain it in detail later on. If, however, the MPUM need not exist, then it is logical to look for the unfalsified model which minimizes the complexity. We will not pursue this point here, but simply give as an example the problem of fitting a polynomial through a given set of data points in $\mathbf{R}^2$. That the polynomial has to pass through these points yields a model (the polynomial) which is unfalsified. To look for the lowest degree polynomial corresponds to minimizing the complexity. In this case we define the complexity of the model to be the degree of the polynomial.

Let us formalize the MPUM for linear, time-invariant, complete dynamical systems. The class of models considered will hence be the (AR) models. Thus each element of the model class M is a linear time-invariant complete dynamical system $\Sigma = (\mathbf{Z}, \mathbf{R}^q, \mathfrak{B})$. The measurements consist of the finite set of time-series $\mathfrak{D}' = \{\tilde{\mathbf{w}}_1, \tilde{\mathbf{w}}_2, \ldots, \tilde{\mathbf{w}}_k\}$ where each $\tilde{\mathbf{w}}_i \colon \mathbf{Z} \to \mathbf{R}^q$ is an observed response. We will call $\Sigma = (\mathbf{Z}, \mathbf{R}^q, \mathfrak{B}) \in M$ *unfalsified* by $\mathfrak{D}$ if $\mathfrak{D} \subseteq \mathfrak{B}$. We will call the element $\Sigma_1 = (\mathbf{Z}, \mathbf{R}^q, \mathfrak{B}_1)$ *more powerful* than $\Sigma_2 = (\mathbf{Z}, \mathbf{R}^q, \mathfrak{B}_2)$ if $\mathfrak{B}_1 \subseteq \mathfrak{B}_2$. We will call $\Sigma^*_{\mathfrak{D}'} = (\mathbf{Z}, \mathbf{R}^q, \mathfrak{B}^*_{\mathfrak{D}'}) \in M$ the *most powerful unfalsified model* in our model class if $\mathfrak{D} \subseteq \mathfrak{B}^*$, and if for any other unfalsified model $\Sigma = (\mathbf{Z}, \mathbf{R}^q, \mathfrak{B}) \in M$ there holds $\mathfrak{B}^*_{\mathfrak{D}'} \subseteq \mathfrak{B}$. It is important to observe the easy but important

PROPOSITION. $\mathfrak{B}^*_{\mathfrak{D}'}$ exists.

In [1] several numerical algorithms are discussed for passing from $\mathfrak{D}'$ to $\mathfrak{B}^*_{\mathfrak{D}'}$. The most powerful unfalsified model $\mathfrak{B}^*_{\mathfrak{D}'}$ can be characterized as the unfalsified model which has the minimal number of input variables and, among these, the minimal number of states.

16. Approximate Modelling

It goes almost without saying that the *real* problem in system identification is to obtain a model of limited complexity which fits the measurements approximately. A direct formulation and algorithms for such an approximation approach, entirely in a deterministic framework, is presented in [1].

The guiding principle which we follow in choosing an element from the model set $\mathfrak{B}$ on the basis of a set of measurements $\mathfrak{D}'$ is that we want a model which has a low complexity (expressing the *a priori* appeal of the model) and low misfit (expressing our faith in the model as put into evidence by the observations). We will now formalize this for models *without* latent variables.

Let $M \subset 2^{\mathfrak{U}}$ be the model class and $D \subset 2^{\mathfrak{U}}$ be the set of possible measurements. Now introduce two partially ordered spaces: $\mathcal{C}$, the *complexity level space,* and $\mathcal{E}$, the *misfit level space,* and two maps, $c \colon M \to \mathcal{C}$, the *complexity,* and $\epsilon \colon D \times M \to \mathcal{E}$, the *misfit.* The complexity is a measure which expresses how complicated we consider a model to be. The more complex a model, the more undesirable it is. As such, we view complexity as being related to how much a model explains. Of course, complexity could also express other features related to the usefulness or the æsthetic appeal of a model. The misfit is a quantitative measure which expresses the degree to which our model is corroborated by the measurements. The larger the misfit, the more we should distrust the model.

Let $M \subset 2^{\mathfrak{U}}$, $D \subset 2^{\mathfrak{U}}$, $c \colon M \to \mathcal{C}$, and $\epsilon \colon D \times M \to \mathcal{E}$ be a modelling set-up as described above. Our modelling algorithms are based on the following procedures.

1. *Modelling with a maximal tolerated misfit.* Let $\epsilon^{\text{tol}} \in \mathfrak{E}$ and $\mathfrak{D}' \in D$ be given. We will call $\mathfrak{B}^* \in M$ the *optimal approximate model within the misfit tolerance* if:

(i) $\epsilon(\mathfrak{D}', \mathfrak{B}^*) \leq \epsilon^{\text{tol}}$;

(ii) $\{\mathfrak{B} \in M, \epsilon(\mathfrak{D}', \mathfrak{B}) \leq \epsilon^{\text{tol}}\} \Rightarrow \{c(\mathfrak{B}^*) \leq c(\mathfrak{B})\}$;

(iii) $\{\mathfrak{B} \in M, \epsilon(\mathfrak{D}', \mathfrak{B}) \leq \epsilon^{\text{tol}}, c(\mathfrak{B}) = c(\mathfrak{B}^*)\} \Rightarrow$
$\{\epsilon(\mathfrak{D}', \mathfrak{B}^*) \leq \epsilon(\mathfrak{D}', \mathfrak{B})\}$.

The interpretation of these conditions should be clear. The first two formalize that $\mathfrak{B}^*$ is the least complex model which fits the model within a given tolerated misfit. If there are more such models which are minimally complex, then (iii) tells us to choose among these the best fitting one.

2. *Modelling with a maximal admissible complexity.* Let $c^{\mathrm{adm}} \in C$ and $\mathfrak{D}' \in D$ be given. We will call $\mathfrak{B}^* \in M$ the *optimal approximate model within the admissible complexity* if:

(i) $c(\mathfrak{B}^*) \leq c^{\mathrm{adm}}$;
(ii) $\{\mathfrak{B} \in M, c(\mathfrak{B}) \leq c^{\mathrm{adm}}\} \Rightarrow \{\epsilon(\mathfrak{D}', \mathfrak{B}^*) \leq \epsilon(\mathfrak{D}', \mathfrak{B})\}$;
(iii) $\{\mathfrak{B} \in M, c(\mathfrak{B}) \leq c^{\mathrm{adm}}, \epsilon(\mathfrak{D}', \mathfrak{B}) = \epsilon(\mathfrak{D}', \mathfrak{B}^*)\} \Rightarrow$
$\{c(\mathfrak{B}^*) \leq c(\mathfrak{B})\}$.

The first two of these conditions show that $\mathfrak{B}^*$ is the best fitting model within the given admissible complexity. If there are more admissible model which achieve this best fit, then (iii) tells us to choose the least complex one.

We will now briefly and informally discuss one version of these algorithms applied to time-series modelling with linear, time-invariant, complete dynamical systems. Let $\tilde{w}: \mathbf{Z} \to \mathbf{R}^q$ be an observed time-series which we would like to model by means of an (AR) model. Assume that, contrary to what is usually done, we specify *a priori* neither the required number of (AR) relations (in the notation of Section 10, g is thus a variable to be determined), nor their lag structure (in particular, the order of underlying state space is also a variable to be determined). However, we will specify the misfit level with which we desire the equations to be satisfied.

In order to define the misfit level, we assume that the observed time series $\tilde{w}: \mathbf{Z} \to \mathbf{R}^q$ belongs to a shift-invariant inner product space $\mathcal{K} \subset (\mathbf{R}^m)^{\mathbf{Z}}$. This condition guarantees, in particular, that for any $a(s) \in \mathbf{R}^{1 \times q}[s]$, $\|a(\sigma)\tilde{w}\|_{\mathcal{K}}$ is a well-defined real number.

Now assume that the linear, time-invariant, complete system $\Sigma = (\mathbf{Z}, \mathbf{R}^q, \mathfrak{B})$ is proposed as a model deduced from the observed $\tilde{w} \in \mathcal{K}$. *How should we define the misfit level $\epsilon(\tilde{w}, \mathfrak{B})$ with which $\tilde{w}$ fails to corroborate $\mathfrak{B}$?* One possibility (the one developed in [1]) is the following. We will define the misfit level $\epsilon = (\epsilon_0, \epsilon_1, \dots, \epsilon_\ell, \dots)$ to be a sequence of real numbers, where $\epsilon_\ell(\tilde{w}, \mathfrak{B})$ will have the intuitive meaning of being the error with which $\tilde{w}$ fails to corroborate the scalar (AR) lag relations induced by $\mathfrak{B}$ and which are 'truly' of order ℓ.

What do we mean by this? The behavior $\mathfrak{B}$ can be expressed as $\mathfrak{B} = \ker R(\sigma, \sigma^{-1})$. Now consider the following subset of $\mathbf{R}^{1 \times q}[s]$, denoted by $\mathfrak{L}$, and defined as $\mathfrak{L} = \{a(s) \in \mathbf{R}^{1 \times q}[s] \,|\, a(\sigma)\mathfrak{B} = 0\}$. This set $\mathfrak{L}$ consists of all scalar (AR) relations satisfied by all time-series which belong to the behavior $\mathfrak{B}$. Clearly $a(s) \in \mathfrak{L}$ implies $p(s)a(s) \in \mathfrak{L}$ for all $p(s) \in \mathbf{R}[s]$. In fact, $\mathfrak{L}$ is the submodule of $\mathbf{R}^{1 \times q}[s]$ generated by the rows of $R(s, s^{-1})$. However, we will not use this module structure.

Now define $\mathfrak{L}_\ell$ as follows:

$$\mathfrak{L}_\ell = \{a(s) \in \mathfrak{L} | \text{ the degree of } a(s) \text{ is } \le \ell\}.$$

Thus $\mathfrak{L}_\ell$ denotes all the scalar (AR) relations of lag less than or equal to ℓ which are satisfied by the behavior $\mathfrak{B}$. Clearly $a(s) \in \mathfrak{L}_\ell$ implies $sa(s) \in \mathfrak{L}_{\ell+1}$. Consequently, in order to specify the scalar (AR) relations which are 'truly' of lag ℓ, we should somehow weed out those which can be obtained from simply multiplying a lower order lag relation by an appropriate polynomial. One way of doing this is by defining $\tilde{\mathfrak{L}}_0 = \mathfrak{L}_0$ and $\tilde{\mathfrak{L}}_\ell = \mathfrak{L}_\ell \cap (\mathfrak{L}_{\ell-1} + s\mathfrak{L}_{\ell-1})^\perp$, where $s\mathfrak{L}_\ell = \{a(s) \in \mathfrak{L}_{\ell+1} | \text{there exists } a'(s) \in \mathfrak{L}_\ell \text{ such that } a(s) = sa'(s)\}$, and the orthogonal complement is obtained by identifying the element $a(s) = a_0 + a_1 s + \ldots + a_\ell s^\ell$ of degree $\le \ell$ in $\mathbf{R}^{1 \times q}[s]$ with the vector $[a_0 | a_1 | \ldots | a_\ell]^T$ and using the standard Euclidean inner product in $\mathbf{R}^{(\ell+1)q}$. Now it is reasonable to view $\tilde{\mathfrak{L}}_\ell$ as the scalar (AR) relations satisfied by $\mathfrak{B}$ and *which are 'truly' of lag* ℓ.

This leads to the following definition of the misfit $\epsilon_\ell(\tilde{\mathbf{w}}, \mathfrak{B})$:

$$\epsilon_\ell(\tilde{\mathbf{w}}, \mathfrak{B}) = \max_{a(s) \in \tilde{\mathfrak{L}}_\ell} \frac{\|a(\sigma)\tilde{\mathbf{w}}\|_\kappa}{\|a\|_\varepsilon}$$

where $\| \cdot \|$ denotes the standard Euclidean norm induced on $\tilde{\mathfrak{L}}_\ell$; if $\tilde{\mathfrak{L}}_\ell = \{0\}$, we put $\epsilon_\ell(\tilde{\mathbf{w}}, \mathfrak{B}) = 0$. With this definition, we obtain a well-defined sequence of nonnegative real numbers

$$(\epsilon_0(\tilde{\mathbf{w}}, \mathfrak{B}), \epsilon_1(\tilde{\mathbf{w}}, \mathfrak{B}), \ldots, \epsilon_\ell(\tilde{\mathbf{w}}, \mathfrak{B}), \ldots) = \epsilon(\tilde{\mathbf{w}}, \mathfrak{B}),$$

at most a finite number of which are nonzero.

Now assume that a maximally tolerated misfit level

$$\epsilon^{\text{tol}} = (\epsilon_0^{\text{tol}}, \epsilon_1^{\text{tol}}, \ldots, \epsilon_\ell^{\text{tol}}, \ldots)$$

is imposed. We then call a model with behavior $\mathfrak{B}$ *tolerated* if the quantity $\epsilon_\ell(\tilde{\mathbf{w}}, \mathfrak{B}) \le \epsilon_\ell^{\text{tol}}$ for $\ell = 0, 1, \ldots$. It is now reasonable to take the

most desirable tolerated model to be the one which maximizes sequen-
tially dim $\tilde{\mathfrak{L}}_0$, dim $\tilde{\mathfrak{L}}_1, \ldots$, dim $\tilde{\mathfrak{L}}_\ell, \ldots$. In [1] we have given a numerical
algorithm based on singular value decomposition which computes such
an 'optimal' approximate model. We refer to this reference for formalization, details, simulations, etc.

Summarizing: We view the problem of constructing a model
from data as a trade-off between the complexity, the misfit, and the
latency. Of particular interest is the most powerful unfalsified model.
For linear, time-invariant, complete systems such a model always exists.

References

[1] Willems, J. C., "From Time Series to Linear Systems: Part I—
Finite-Dimensional Linear Time-Invariant Systems; Part II—Exact
Modelling; Part III—Approximate Modelling," *Automatica,* 22 (1986),
561–580 (Part I), 22 (1986), 675–694 (Part II), 23 (1987), 87–115 (Part
III).

[2] Willems, J. C., "Models for Dynamics," *Dynamics Reported,* Vol. 2,
pp. 171–269, Wiley, Chichester, 1989.

Force, Measurement, and Life

MICHAEL CONRAD

Abstract

It is proposed that the forces between manifest particles are mediated
by chains of virtual pair production processes in a plenum of unmanifest
vacuum fermions. The world lines of both virtual and real bosons are
identified with such sequences, each step coupled to the next by the requirement for macroscopic conservation of energy and momentum. If the
short-lived pairs are identified with positronium in the case of electromagnetic force, the coupling strength is controlled by the fine structure
constant. Application of first-order perturbation (or scattering) theory
yields an inverse-square law incorporating quantized space and time intervals that reflect the density of vacuum fermions. The occurrence of
these intervals eliminates infinite self-energies and allows for a number
of broken symmetries that yield force laws with an apparent noninverse
character (aside from the noninverse square character that arises from
relativistic transformations). An important feature of the model is that
positive energy mass and charge depress the surrounding vacuum density.
This means that propagating pair formation processes must be accompanied by compression-expansion waves of vacuum density, yielding a
wave aspect of radiation. The depression of vacuum density produced
by mass may be identified with space-curvature, and it is shown that
the model incorporates Mach's principle and the principle of equivalence. Strictly speaking, vacuum particles cannot have mass since they
are not absorbers; however, they can inherit mass by acting as transient absorbers. These masses and the structure of the vacuum density
induced by the distribution of positive energy mass and charge must
satisfy self-consistent equilibrium relations. The time evolution of any
system in which these relations are disturbed must exhibit fundamental
irreversibility due to the fact that the classical specification of force depends on the vacuum density structure, hence on the wave function of the
vacuum. This underlying irreversibility is negligible in simple physical
systems, but would be unmasked by the sensitive dynamics of biological
systems and measurement instrumentation.

1. Introduction

Quantum mechanics and quantum field theory have allowed physicists
to construct impressively useful models of matter. For some scientists, however, the methodology for creating these models has always
raised serious doubts about any claim to universality. The problem
is that two distinct types of influence are at work. The first is connected with forces mediated by the exchange of virtual particles and
the second with measurement. The key point is that the modeling

methodology provided by the standard quantum mechanical equations of motion does not extend in a natural way to the process of converting the unpicturable wave function whose time evolution they describe to a picturable form that can be represented in the pointer reading of a measuring instrument [1–4]. The methodology thus fails in a conceptual way just for the type of process that we expect humans and other biological systems to perform.

We can fairly easily trace the historical origin of this conceptual dissonance. In the physics of Aristotle the motion of objects was controlled by basins of attraction. As a consequence, force, or influence, had to be irreversible (for example of the form $F \propto dx/dt$). The main point of Newtonian physics was to remove this irreversibility as a fundamental feature. Thus the ubiquitous procedure for modeling a system was to suppose that the change in the change produced by any influence is constant (in a manner expressed by the form $F \propto d^2x/dt^2$). The manifest irreversibility and indeterminism of macroscopic experience had to be assigned a fictitious status, raising the new problem of passing from the reversible to the irreversible and from the determinate to the indeterminate. The equations of motion of quantum mechanics and quantum field theories retain the second derivative form of Newton's equation, hence the fundamental feature of time reversibility. The difference is that what develops reversibly in time is not a definite state, with definite numbers assigned to classical variables, but a wave function representing a superposition of possibilities. This required the introduction of a new type of influence, connected with the process of collapsing the unpicturable wave function into a classical state in the measurement process. As a consequence irreversibility and probability returned in a truly fundamental form that could not simply be attributed to the subjective ignorance of the observer.

Suppose that the process of wave function collapse could in principle be reduced to a reversible equation of motion. This would mean that the wave function could in principle be collapsed into a classical state in a reversible manner and without probability generation, apart from the purely subjective lack of knowledge on the part of the observer. But this in turn would mean that there exists in principle a complete set of precise initial conditions for a microphysical system, in violation of the uncertainty principle. This is conceivable, but only if one admits a more fundamental equation of motion, formulated in terms of hidden variables and (according to Bell's theorem) incorporating action at a distance [5].

The purpose of this paper is to consider whether an alternative picture is possible that accommodates both the conventional forces (electromagnetic, gravitational, weak, strong) and the process of measurement. The idea is that the type of force that enters into equations of motion and the phenomenon of wave function collapse are both special cases of a broader process in a manner reminiscent of a broken symmetry. In this broader process the interactions among particles in general lead to wave function collapse and irreversible time development. But in the simple systems ordinarily studied in the physics laboratory this irreversibility is either quenched or too negligible to detect. The prime feature of measuring instruments and biological systems is that they unmask the latent irreversibility.

The underlying idea is that the interactions between particles depends on the structure of the surrounding "vacuum." Here we can view the vacuum as a plenum of negative energy fermions, just in the sense of a Dirac vacuum [6]. The virtual bosons that mediate the conventional forces are propagating excitations of these negative energy fermions, specifically transient pair production processes that trigger one another in sequence (in the fashion of a skipping process). As long as the motion of the interacting particles does not disturb the vacuum structure, its wave function remains the same. When the vacuum structure is altered, however, the classical description of any force mediated by it requires a collapse (or projection) process to occur, revealing the underlying irreversibility. Our hypothesis is that the regime of collapse and collapse-associated dissipation dominated the early universe, and that biological systems are a return to this regime on a smaller but more controlled scale.

We will develop the model in three stages. The first (Section 2) will deal with the quantization of vacuum particle exchanges per se. We will show how a variety of force laws can be expressed in terms of time and space intervals associated with vacuum particle fluctuations and we will describe the mechanism of propagation of these excitations. In the second (Section 3) we will introduce the possible physical interpretations, and will present arguments to the effect that excitations involving transient pair production processes are physically tenable. In this second stage we will briefly consider how the vacuum particle model could in principle accommodate general relativity, but without pursuing this direction in detail here. In the third stage (Section 4) we will return to the issue of measurement, and, in our concluding remarks, to the implications for life and mind.

What is to be presented is a model, not a thoroughgoing quantum mechanical theory. The objective is to consider whether a skipping mechanism of virtual particle exchange can yield force laws and at the same time additional features that should be present in a fundamental theory.

2. Quantization of Virtual Particle Exchanges

2.1. The skipping model. Quantum field theories provide a mechanistic explanation of force in terms of the exchange of virtual particles. Such particles are usually viewed as field quanta whose existence violates conservation of energy (as allowed by the time-energy uncertainty principle). In covariant treatments energy is usually considered to be conserved, with the virtual particles existing by virtue of a transient suspension of the relativistic energy-momentum relation. In either interpretation the exchange is pictured as a continuous process potentially involving a continuous range of energies and momenta (aside from restrictions connected with renormalization). The potential function of the field (for example, the inverse square field in the case of massless momentum carriers) can be viewed as controlled by the wave function of the Bose particles which mediate the field [7].

We need to consider whether an alternative picture of the exchange process is possible in which the exchange is interpreted as occurring in a series of steps rather than as a continuous process. As already indicated the virtual particle in this modified picture is identified with a propagating chain of transient excitations, called a skipping process. The energies and momenta which may be possessed by these constituent transient processes are highly restricted. The potential function of the field (such as the inverse square field) is connected with the spatial distribution of such highly localized virtual processes rather than with a distribution of energies and momenta of carrier particles continuously existing during the exchange.

To gain a foothold on this idea, we first present a heuristic argument which suggests the existence of a skipping process. The argument is based on a derivation of the inverse square law from first order, time-independent perturbation theory. The derivation requires the assumption that the relation between the distance separating two observable (long-lived) particles and the energy involved in coupling them by a series of transient (virtual) processes is linear. This derivation has the aspect of a formal trick. But since it leads to a result with the correct form it seems worthwhile to consider its implications and to ask whether it can be justified on the basis of these implications.

The main implication is that the time and space intervals characteristic of the transient processes which mediate the force are quantized. These characteristic intervals are called the fluctuation time and fluctuation volume, respectively. As the density of transient processes (to be called the fluctuation density) increases, the strength of the force mediated by these transient processes decreases. The occurrence of such localized virtual excitations is due to a coupling of short-lived violations of energy and momentum conservation. It is not possible for these annihilations to disappear simultaneously in the absence of mass. As a consequence they must propagate. This is the essence of the skipping model.

The dynamics of skipping is analyzed in a manner which is independent of physical interpretation. But for quantized force laws and the skipping process which they imply to be physically tenable it is clearly necessary that at least one physically consistent interpretation can be constructed. The suggestion to be made is that each skip corresponds to a transient excitation of a vacuum fermion and that the wave properties of boson fields are connected with alterations of vacuum density produced by mass and charge. Real bosons may be interpreted as skipping processes in which the accompanying vacuum density wave carries real energy.

The initial analysis is nonrelativistic since the derivation of the inverse square law employs the time-energy uncertainty principle and therefore most naturally uses noncovariant perturbation theory. However, we show that the constancy of light velocity can be treated as a thermodynamic consequence of the existence of fluctuation quantities and that fluctuation time and fluctuation length are intrinsically Lorentz invariant. The quantized inverse square law allows for a number of forms of symmetry breaking which yield force laws with an apparent noninverse square character.

2.2. Perturbation theory approach to the inverse square law. According to first order, time dependent perturbation theory, the probability that a system undergoes a transition from state i to state f in a time interval $\tau = t$ in the presence of a perturbing potential constant over that time interval is given by

$$p(i \rightarrow f) = \frac{4|<f|V|i>|^2 \sin^2\{\frac{1}{2}(\Delta E/\hbar)\tau\}}{\Delta E^2} \tag{1}$$

where $\Delta E = E_i - E_f$. The probability of undergoing a transition increases with τ, in accordance with intuition. However, as τ increases, the probability of undergoing a transition to an equal energy state increases as τ^2, consistent with the fact that conservation of energy should hold rigorously over macroscopic times [6, 8]. This τ^2 increase in the probability of satisfying energy conservation is a key feature of the model to be developed. The intuitive idea is that the probability that a virtual momentum carrier satisfies energy conservation by actually effectuating a transfer of momentum increases as the square of τ, interpreted as a characteristic of the vacuum structure. τ will be called the fluctuation time and the perturbation V will be viewed as a self-perturbation allowed by a virtual fluctuation of energy compatible with $\sigma E \sigma \tau \approx \hbar$, where $\sigma \tau \approx \tau$. (Values of σE much larger than $\hbar/\tau$ may be ignored since the fluctuations represent violations of a conservation law, not all the uncertainty accumulated in an experimental measurement.) Taken in the context of scattering theory Eq. (1) is the same as the Born approximation, though again it should be emphasized that in scattering theory V is usually taken as the potential which the present model is intended to derive [9].

Consider two particles separated by a distance r and exchanging momentum through a virtual process. The tentative assumption will be that the energy exchange which must occur over any given interval of time in order to achieve any given degree of coupling increases linearly with the distance. The simplest linear dependence is given by

$$\Delta E_{\text{req}} = Kr \tag{2}$$

where K is a proportionality constant with dimensions of force. Of course the energy exchanged must actually decrease with distance, but this is consistent with the fact that the degree of coupling also decreases. If $d^2(\Delta E_{\text{req}})/dr^2 \neq 0$, the momentum exchanged over any given time interval to achieve a given degree of coupling would either have to increase or decrease with r. But if the relation $E^2 = p^2c^2 + m^2c^4$ is not detectably violated by virtual particles the degree of coupling should depend on the momentum exchanged alone. Thus linearity is the simplest assumption consistent with conservation of momentum.

According to the time-energy uncertainty principle

$$\Delta E \approx \frac{\hbar}{\tau_{\Delta E}} \tag{3}$$

The fluctuation magnitude ΔE corresponds to the standard deviation of the energies σE, which would be obtained by measurements made at some point in the time interval $\tau_{\Delta E}$. $\tau_{\Delta E}$ is the longest fluctuation time compatible with a fluctuation magnitude ΔE, and may be interpreted as the standard deviation of the times at which the measurement is made. Substituting any choice of $\tau_{\Delta E} = \tau$ in Eq. (3) into the numerator of Eq. (1) gives

$$p(i \to f) \approx \frac{0.92| < f\,|V|\,i > |^2}{\Delta E^2} \qquad (4)$$

For values of $\Delta E \approx \Delta E_{\text{req}}$ we can also write

$$p(i \to f) \approx \frac{0.92| < f\,|V|\,i > |^2}{K^2 r^2} \qquad (5)$$

The force should be proportional to the probability if the only influence the particles can exert on each other is through the exchange of momentum. Thus for an ensemble of exchanges

$$\mathbf{F} \approx \frac{\hat{A}}{K^2 r^2} \frac{\mathbf{r}}{r} \qquad (6)$$

where $\hat{A} = A| < f\,|V|\,i > |^2$ and A is the proportionality constant. The appearance of r^2 in the denominator shows that the inverse square law is simply an expression of the linearity relation (2), independent of the details of the matrix elements. It should be emphasized that V is not the potential function used to calculate, say, the Coloumb force in field theory, but rather the fluctuation quantity $\hbar/\tau_{\Delta E}$ itself (soon to be associated with each of a sequence of transient processes which mediate the force).

2.3. Fluctuation times and fluctuation lengths. Since the linearity assumption appears to be justified in at least some important cases, it is interesting to take it as a postulate and investigate its consequences. Linearity can be written as

$$\Delta E_{\text{req}} = K c \tau_{\Delta E} \qquad (7)$$

where c is the velocity of light and $r = c\tau_{\Delta E}$. But in each case $\Delta E_{\text{req}} \Delta \tau_{\Delta E} \approx \hbar$. Eliminating ΔE_{req} from (7) yields

$$\tau_{\Delta E}^2 \approx \frac{\hbar}{Kc} \qquad (8)$$

Thus the fluctuation time, or at least its minimum value, is fixed by K, independent of the virtual fluctuation in energy ΔE_{req}. To express this parametric dependence on K we will from now on write $\tau_{\Delta E}$ as τ_K. Substituting into the force formula (Eq. 6)

$$\mathbf{F} \approx \frac{\hat{A} c^2 \tau_K^4}{\hbar^2 r^2} \frac{\mathbf{r}}{r} \tag{9}$$

But r can be written as

$$r = N r_K \tag{10}$$

where $r_K^2 = x_K^2 + y_K^2 + z_K^2$ and N is the number of fluctuations separating the two particles. Similarly,

$$r = N c \tau_K \tag{11}$$

where $r_K = c \tau_K$. Thus

$$\mathbf{F} \approx \frac{\hat{A} \tau_K^2}{\hbar^2 N^2} \frac{\mathbf{r}}{r} \tag{12}$$

Note that the real structure of this argument is that for individual exchanges, the linearity assumption and the uncertainty principle are incompatible except for the fixed distance r_K and for $N = 1$. When $N > 1$ the chance that the propagating chain initiated by one absorbing particle is absorbed by a second one decreases with the distance between them, therefore as $1/N^2$ (see Section 4). As a consequence, ΔE_{req} must in general represent the fluctuation energy of an individual exchange multiplied by the number of chains required for a given coupling strength, which is the reciprocal of the chance that the chain will actually be absorbed.

We are now in a position to consider the justification for using first-order perturbation theory. Eq. (5) and all of the force expressions derived from it hold if second-order, time-dependent perturbation theory is taken as a starting point, except that $|<f|V|i>|^2$ would have to be replaced by $|<f|V|i> - \sum_{u \neq f,i} <f|V|u><u|V|i>|^2$. Here u is an intermediate state. However, the τ_K^2 dependence suggests that such second-order processes are not important and Eq. (7) implies that they are incompatible with the linearity assumption. The reason is that the occurrence of an intermediate state would require the fluctuation time to be decomposed into constituent fluctuation times which are smaller. Processes occurring on such smaller time scales occur, but do not make a significant contribution to the coupling between two observable particles. As a consequence the exchange of momentum between the particles is properly treated as consisting of $N = r/r_K$

first-order processes best described by the first order matrix element $< f \, |V| \, i >$.

It is interesting to consider the relation between the fluctuation time and the number of fluctuations. As the fluctuation time increases, the maximum amount of momentum which can be exchanged in each fluctuation decreases since the fluctuation energy ΔE, must decrease. Nevertheless the resulting force becomes larger according to Eq. (12). For this to be possible, the number of fluctuations per unit time which actually effect a momentum transfer must increase fast enough to compensate the decrease in the momentum exchanged. This fast increase is ultimately based on the fact that the probability of violating energy conservation decreases as τ^2 in the original perturbation formula. As a consequence the probability that a virtual particle is exchanged increases as the square of the fluctuation time (as in Eq. 12), whereas the maximum energy which it is likely to exhibit decreases as the inverse power of fluctuation time. *Thus, as the minimum time required for a virtual process to effectuate an exchange of momentum increases, the number of such processes which actually do lead to an exchange of momentum also increases.* In the next section we shall see that this time is a characteristic of the vacuum structure.

A similar conclusion holds for fluctuation lengths. Substituting $r_K^2 = c^2 \tau_K^2$ into Eq. (9)

$$F \approx \frac{\hat{A} r_K^4}{c^2 \hbar^2 r^2} \frac{\mathbf{r}}{r} \qquad (13)$$

Using Eq. (10) to eliminate explicit reference to r

$$F \approx \frac{\hat{A} r_K^2}{c^2 \hbar^2 N^2} \frac{\mathbf{r}}{r} \qquad (14)$$

When the fluctuation length increases, the fluctuation in momentum should decrease as $\Delta p_x \approx \hbar/x_K$ and similarly for the other components of momentum. *Thus as the minimum space interval which a virtual particle must traverse in order to effectuate an exchange of momentum increases, the fraction of virtual events which actually lead to an exchange of momentum also increases.*

2.4. Fluctuation densities and field strengths. It is useful to express F in terms of a density of events in the vacuum. Since the number of events per unit volume is given by $\rho_K = 1/r_K^3$, Eq. (13) can be written as

$$F \approx \frac{\hat{A}}{c^2 \hbar^2 \rho_K^{\frac{4}{3}} r^2} \frac{\mathbf{r}}{r} \qquad (15)$$

Equating $\mathbf{F}$ in Eq. (15) with $\mathbf{F}_{grav}$ in the case of two interacting objects of mass m,

$$\mathbf{F}_{grav} = -\frac{Gm^2}{r^2}\frac{\mathbf{r}}{r} \approx \frac{\hat{A}_{grav}}{c^2\hbar^2\rho_{grav}^{\frac{4}{3}}r^2}\frac{\mathbf{r}}{r} \qquad (16)$$

where G is the gravitational constant, ρ_{grav} is the fluctuation density for the gravitational field, $\hat{A}_{grav} = A_{grav}| < f|V_{grav}|i > |^2$, and we take A_{grav} as negative. Eliminating r gives

$$m \approx \pm\frac{\hat{A}_{grav}^{\frac{1}{2}}}{G^{\frac{1}{2}}c\hbar\rho_{grav}^{\frac{2}{3}}} \qquad (17)$$

Similarly equating $\mathbf{F}$ with $\mathbf{F}_{Coul}$ for two interacting objects each carrying charge q, we obtain

$$\mathbf{F}_{Coul} = \frac{q^2}{r^2}\frac{\mathbf{r}}{r} \approx \frac{\hat{A}_{em}}{c^2\hbar^2\rho_{em}^{\frac{4}{3}}r^2}\frac{\mathbf{r}}{r} \qquad (18)$$

where ρ_{em} is the fluctuation density for the electromagnetic field and $\hat{A}_{em} = A_{em}| < f|V_{em}|i > |^2$. Eliminating r gives

$$q \approx \pm\frac{\hat{A}_{em}^{\frac{1}{2}}}{c\hbar\rho_{em}^{\frac{2}{3}}} \qquad (19)$$

Thus as mass increases, the gravitational fluctuation density decreases. As charge increases, the electromagnetic fluctuation density decreases.

Since the gravitational force is weaker than the electromagnetic force, Eq. (15) implies that ρ_{grav} must be larger than ρ_{em}. The fluctuation times and fluctuation volumes must therefore be smaller for the vacuum processes which mediate the gravitational force than they are for the processes which mediate the electromagnetic force. Since the quantity $F_{grav}/F_{Coul} \sim 10^{-43}$ for the forces between two electrons, the density of events in the surrounding gravitational field as compared to the surrounding electromagnetic field is approximately $\rho_{grav}/\rho_{Coul} \sim 10^{32}$. This means that the number of events which contribute to the force between two electrons is on the order 10^{11} greater for the weak gravitational interaction than for the stronger electromagnetic one. For two protons the number of gravitational events is about 10^{10} greater since F_{grav}/F_{Coul} is about 10^{37} in this case. This is due to the altering effect of mass on vacuum density featured in Eq. (17).

2.5. Consistency with special relativity. The starting assumptions from which the concepts of fluctuation time, fluctuation length, and fluctuation density have been developed are valid in a three space. The inverse square law used to exhibit these quantities is by itself noncovariant. Nevertheless, two considerations suggest that the existence of fluctuation quantities is compatible with a relativistic framework. The first is that the constancy of light velocity (assumed in the relation $r_K^2 = c^2 \tau_K^2$) can be deduced as a thermodynamic requirement imposed by the linearity assumption. This argument is presented since it suggests that the four space geometry of special relativity can be viewed as a macroscopic property of space-time which is completely consistent and even implied by the validity of three space uncertainty principles in the small. A second consideration is that fluctuation time and length are themselves Lorentz invariant, as shown in the next section.

First we consider the consistency of the assumption that the fluctuation length r_K is related to the fluctuation time by $r_K^2 = c^2 \tau_K^2$. Since τ_K is the time interval during which the virtual fluctuation of energy can exist, r_K may be interpreted as the space interval over which it can extend. According to the position-momentum uncertainty principle

$$\Delta p_x x_K \approx \hbar \tag{20}$$

where Δp_x is the momentum fluctuation corresponding to the energy fluctuation ΔE. As with the time-energy uncertainty relation (Eq. 3), the assumption of a near equality is appropriate since the momentum violation represents a violation of a conservation law rather than all the accumulated uncertainty of an experimental measurement. Equating these two uncertainty relations

$$\Delta E \tau_K \approx \Delta p_x x_K \tag{21}$$

Since this relation holds for all components of momentum we can write

$$\Delta E \approx \Delta pc \tag{22}$$

The relation between the fluctuation energy and the fluctuation momentum contributing to the inverse square force between two particles thus corresponds to the relation between the energy and momentum of a massless particle, as is reasonable. However ΔE and Δp have the important feature that they can only assume the values $\hbar/\tau_K$ and $\hbar/c\tau_K$, respectively.

Suppose that c is not constant in all inertial systems. The relation between fluctuation length and fluctuation time would then be $x_K = v_x \tau_K$ (for simplicity we consider only the x component of velocity). Eq. (22) becomes

$$\Delta E \approx \Delta p_x v_x \qquad (23)$$

This contradicts the linearity assumption since the momentum exchanged would then depend on the velocity of the coordinate system. Replacing c by v_x in the linearity relation Eq. (7), the force law in terms of fluctuation times (Eq. (9)) is replaced by

$$F_x \approx \frac{\hat{A} v_x^2 \tau_K^4}{\hbar^2 \tau^2} \qquad (24)$$

The thermodynamically unacceptable feature is the occurrence of a velocity in an inverse square law (other than the constant c). If we transform to a coordinate system in which v_x is smaller, the force between two particles would decrease even though the inverse square character of the law remains intact.

Suppose that a pulley used for lifting or dropping a weight is attached to an elevator which travels in a vertical shaft perpendicular to the earth. The elevator serves as the coordinate system for the observer, who may be thought of as the pulley operator. The elevator may have two velocities, V_x^{up} and V_x^{down}, with the convention $V_x^{up} > V_x^{down}$. To eliminate the complication of an accelerating coordinate system we suppose that the elevator is not allowed to fall freely (for example, it may be required to move in steps, each of equal length, but short enough so that the velocity during each ascending or descending step is approximately constant). Though not necessary for the argument, we also make the simplifying assumption that the weight moves in a series of steps, so that the velocity with which it rises or falls is approximately constant.

The reductio assumption is that v_x, the velocity of the momentum exchange between earth and weight, is smaller in one of the two coordinate systems. We suppose that it is largest when the elevator and weight both move toward the earth and smallest when both are rising (the argument could also be constructed on the basis of the opposite assumption). The operator is free to lift the weight when he is traveling up in the elevator and to let it fall when he is traveling down. According to Eq. (24) the work performed by the falling weight will be greater on the downward trip (when v_x is largest) than the work required to lift it on the upward trip (when v_x is smallest). If the weight is sufficiently large the pulley operator could use this difference in energy to relift the lowered weight and to raise the elevator, thereby creating a perpetual motion device. The reductio assumption that v_x can be different in different inertial coordinate systems must therefore be rejected, implying that the constancy of light velocity in free space can be viewed as being imposed by the linearity requirement rather than as a separate assumption.

Note that in the above argument we have only considered values of v_x greater than or equal to zero. As v_x becomes large and negative, F_x increases according to Eq. (24), but even under the assumption of variable v_x the force would have to be zero in this case since no exchange would occur. Also note that the argument assumes that the fluctuation quantities are constant. But Eqs. (17) and (19) suggest that they can be altered by mass and charge (these effects are discussed in Section 3). The fluctuation quantities remain constant in the relativistic description constructed in the next section, but the force does not retain its inverse square character.

2.6. Incorporation of special relativity. To construct a covariant description, consider two observers L and L^*, moving with velocity v relative to one another along the x-axis. It is not a possible experiment for either L or L^* to measure fluctuation length or fluctuation time either in their own coordinate system or in the others. If these quantities could be observed directly, the virtual violation of energy, ΔE, could be observed directly and used to perform work, in violation of macroscopic thermodynamics. However the following argument shows that r_K and τ_K should be identical in all inertial coordinate systems.

First consider the proper time interval

$$\hat{\tau}_K^2 = \tau_K^2 - \frac{r_K^2}{c^2} \tag{25}$$

where the invariant $\hat{\tau}_K$ is the proper fluctuation time. If the virtual particle emitted travels at the speed of light (here we assume it is a photon or a graviton), the proper time interval between emission and absorption must be zero, giving

$$r_K^2 = c^2 \tau_K^2 \tag{26}$$

which is identical to the definition of fluctuation length used earlier. Clearly the fluctuation length must have the same transformation properties as the fluctuation time, a requirement which also follows from the fact that the force equation expressed in terms of τ_K (Eq. 12) has the same form as the force equation expressed in terms of r_K (Eq. 14). If τ_K and r_K did not transform the same way, an observer calculating the force seen by a second observer moving at a constant velocity with respect to him would arrive at different results depending on which one of these equations he chose to use.

A second consideration is that the equation $r^2 = N r_K^2 = N^2(x_K^2 + y_K^2 + z_K^2)$ implies that the number of fluctuations is the same for any observable component of the fluctuation length (that is, for any component not equal to zero). This is a necessary condition for a single propagating sequence of fluctuations. Suppose that the state of motion of L and L^* differ only along the x-axis. Then N must be the same for y_K^* and z_K^* as for y_K and z_K. But N is the same for y_K^* as for x_K^*, therefore it must be the same for x_K and x_K^*. Yet in an ensemble of experiments the distribution of virtual particles emitted should appear to form a spherical wave from the point of view of every inertial observer. If N is constant this requirement can only be met if r_K and τ_K are Lorentz invariant.

The invariance of r_K implies that a test particle will appear to undergo a rotation with respect to a source particle, as seen by L^*. If $r_K = r_K^*$ and N is constant we can write

$$r = N r_K = N r_K^* = r^* \tag{27}$$

But r should undergo a Lorentz contraction in the x direction, giving

$$r^{*2} = \left(1 - \frac{v^2}{c^2}\right) x^2 + y^{*2} + z^{*2} \tag{28}$$

For simplicity taking $y = z = 0$ in L, Eq. (27) can hold only if

$$y^{*2} + z^{*2} = \frac{v^2}{c^2} r^2 \tag{29}$$

Thus L^* sees components in the y and z directions. Since $y^* = z^*$ by symmetry and since y^* and z^* Lorentz transform to y and z, the sphere undergoes a rotation x through an angle $\sin^{-1}(v/c)$ as seen from L^*.

There is an interesting connection between this rotation and Terrel's analysis of the transformation properties of geometric objects [10–12]. Suppose that the field of the source particle is modeled by a luminiferous sphere with each point on the sphere representing a potential location of the test particle. Such a sphere will be seen as a sphere by all inertial observers, except for a rotation through an angle equal to the angle computed above. This is due to the fact that the Lorentz contraction is balanced by the extra time required for light to reach an observer from the more distant part of the sphere, so that the only change which could be observed is a rotation. Since the transmission of analog signals provides a proper way for L and L^* to compare their observations on the propagation of the field, it would be inconsistent if the rotation of a sphere as measured by L^* did not agree with the rotation of the field structure derived here, implying the existence of an invariant such as r_K.

The occurrence of this rotation shows that the force law implied by first order perturbation theory and the linearity relation is only an inverse square law when the two interacting particles are stationary. If the interacting particles are not at rest in L^* they will obey the force law obtained by transforming them to L^* from a frame in which they are at rest. As is well known, it is possible to use the above procedure to derive magnetic forces and more generally Maxwell's equations from Coulomb's law [13–15]. Radiation emitted by accelerating particles may be interpreted as propagating chains of transient events which have lost their connection to the source particle. But it should be emphasized that further assumptions are required to derive Maxwell's equations, including the assumption of charge conservation and some assumption limiting the effects of a moving source charge. The important point here is that the concepts of fluctuation time and fluctuation length are compatible with special relativity and in the case of moving charges allow for force laws more general than the inverse square law.

2.7. Noninverse square laws. Of the fundamental forces of nature, only the electromagnetic and gravitational appear to obey a simple inverse square law in the nonrelativistic limit. Furthermore, the relativistic description of the gravitational field in terms of arbitrary coordinate transformations gives it a second rank tensor character, different than the vector character of the electromagnetic field. It is therefore useful to classify those features of the fluctuation density force law (Eq. 15) which allow fields to exhibit different relativistic transformation properties or different spatial dependencies in the nonrelativistic limit.

2.7.1. Inhomogeneity of vacuum density. The inverse square law will have an apparent noninverse square character if the fluctuation density varies with the spatial coordinate r. For example, consider the Yukawa potential, $V(r) = -(q^2/r)e^{-r/R}$, where R is the observed range of nuclear forces, related to the meson rest mass by $R = m_\pi c/\hbar$.

Working in fluctuation lengths (Eq. 13)

$$\frac{\hat{A}r_K^4}{c^2\hbar^2 r^2} \approx -q^2 e^{-r/R}\left(\frac{1}{r^2} + \frac{1}{rR}\right) \tag{30}$$

Taking $\hat{A}/c^2\hbar^2 \propto q^2$,

$$r_K \propto \left[e^{-r/R}\left(\frac{r}{R} + 1\right)\right]^{\frac{1}{4}} \tag{31}$$

as opposed to $r_K \propto (1)^{\frac{1}{4}}$ for an unmasked inverse square field. The Yukawa potential may thus be interpreted as an inverse square law whose inverse square character is hidden by a vacuum inhomogeneity in which the fluctuation length (or fluctuation time) falls off slowly up to a distance R, then falls off rapidly. In an atomic nucleus the vacuum may be structured in a much more complicated way, masking not only the inverse square law but the Yukawa potential as well. (The argument given in Section 2.5 that the momentum carrier must travel with velocity c in all inertial coordinate systems does not apply here since the fluctuation quantities are altered by the presence of mass.)

The opposite type of vacuum inhomogeneity is a bubble of high fluctuation density. Two particles within this bubble would exert very weak forces on each other, but would exert stronger forces on particles outside the bubble. For example, an electron surrounded by such a bubble would exert normal electromagnetic forces on other charged particles outside the bubble, but would exert a weak force on charged particles inside the bubble. The skipping model thus suggests that the weak force is related to the electromagnetic force by a broken symmetry in ρ_{em}. The natural suggestion is that this broken symmetry is the physical mechanism of symmetry breaking responsible for relating the weak and electromagnetic forces in the Weinberg-Salam theory [16].

2.7.2. Stretching and compression of fluctuation lengths. If the fluctuation length (or fluctuation time) increases as two attractively interacting particles are separated, the attraction will increase with distance (cf. Eq. 13). As $r_K \rightarrow \infty, F_{attractive} \rightarrow \infty$, so that the two particles could never be completely separated. On the other side, as $r_K \rightarrow 0, F_{attractive} \rightarrow 0$, so that as the particles come closer together they behave more like free particles. The natural suggestion is that this is the mechanism of quark confinement in the skipping model.

2.7.3. Alteration of surrounding vacuum density. When the fluctuation density ρ_{grav} is identified with the density of vacuum particles, mass may be identified with depressions of vacuum density. These depressions produce asymmetries in the origin and annihilation of skipping processes. Later (in Section 3.8) we will show that the gravitational force between masses can be interpreted in terms of the resulting asymmetries in the distribution of skipping processes impinging on them.

2.8. Nonexistence of self-energies for point particles. The fluctuation quantity formulation of the inverse-square law differs from the classical formulation in that it excludes infinite self-energies and indeed excludes self-energies altogether for a single particle. The classical self-energy for a sphere of uniform surface distribution is $E_{self} = q^2/r$ (the situation is changed only by an increase in the proportionality constant if the volume distribution is uniform [17]). The self-energy in terms of fluctuation quantities may be obtained directly from this and from the fluctuation density expression for charge (Eq. 19)

$$E_{self} \approx \frac{\hat{A}_{em}}{c^2 \hbar^2 \rho_{em}^{\frac{4}{3}} N_{em} r_{em}} \tag{32}$$

where for clarity N is subscripted by the field with which it is associated. In terms of fluctuation distance and numbers

$$E_{self} \approx \frac{\hat{A}_{em} r_{em}^3}{c^2 \hbar^2 N_{em}} \tag{33}$$

In the classical expression there is no intrinsic feature of the formula which prevents r going to zero, leading to an infinite self energy for a point particle. However, in the fluctuation density framework no interaction can occur at all unless $N_{em} \geq 1$. The self-energy can go to infinity only if r_{em} goes to infinity. But in that case the interaction would have to involve an infinite volume of space. If r_{em} were identified with, say, the radius of the electron and allowed to go to zero the self-energy would go to zero as well.

The term self-interaction is strictly speaking inappropriate to the fluctuation density framework. The true self-interaction is a noninteraction corresponding to the excluded case of $N_{em} = 0$. For $N_{em} > 0$, at least one fluctuation length must be involved. For a particle such as the electron there are only two possibilities. The fluctuation length could be smaller than the (finite) radius of the electron, in which case the self-energy of the electron would be finite and small. Or the fluctuation length could be greater than the (finite) radius of the electron, in which case the interaction no longer involves a single point particle. In Section 2.9 we show that if $N_{em} = 1$ the interaction must involve the participation of at least two electrons and that the appearance of a self-interaction is connected with the impossibility of identifying the source of a virtual momentum carrier.

Clearly the self-energy equation in terms of fluctuation densities (Eq. 32) must have the same properties as the self-energy equation in terms of fluctuation lengths and fluctuation numbers (Eq. 33). Suppose that a particle is identified with a sphere of zero fluctuation density, that is, ρ_K equals zero inside the sphere and ρ_K equals ρ_{em} or ρ_{grav} outside the sphere. In this case r_K is the diameter of the sphere. If $r = N_K r_K = N_{em} r_{em}$, Eq. (32) may appear to suggest that the self-energy goes to infinity inside the sphere, and goes to infinity especially rapidly as r_{em} goes to zero. However, the condition $N_{em} \geq 1$ is incompatible with any interactions taking place inside this sphere. The sphere is a region of "vacuumized vacuum" and therefore does not contain any emitters or absorbers of virtual particles within it. According to the fluctuation density formalism no force can originate inside such a region, corresponding to the fact that no force exists when $N_K = 0$.

The absence of an infinite self-energy in the fluctuation density formulation of force laws has its fundamental basis in the quantization of the energy and momentum exchanged in a virtual process. For a given fluctuation time τ_K, the fluctuation energy is a restricted quantity $\Delta E \approx \hbar/\tau_K$. For a given component of the fluctuation length the momentum which may be carried between two particles is a restricted quantity $\Delta p_i \approx \hbar/i_K$, $i = x, y, z$. These restrictions provide a natural cutoff for the momentum carried by virtual particles mediating a force. Fluctuations which do not match the fluctuation density do not contribute to force. The limitation on the momentum which a particle may possess is of obvious interest insofar as the divergences in quantum electrodynamics derive from the absence of such a limit. The momentum exchanged can only go to infinity in the limit of an infinite fluctuation density, that is, when r_K and τ_K go to zero.

The foregoing conclusions extend to relativistic forces. According to Eq. (33) the self-energy of a spherically distributed system of particles can go to infinity only if $r_K^* \to \infty$. But r_K^* is always finite if r_K is (since $r_K^* = r_K$). Thus if the self-energy is finite in a coordinate system in which the inverse square law holds it must be finite in any other inertial coordinate system.

2.9. Skipping mechanism. The quantization of fluctuation time and fluctuation length is incompatible with a purely continuous picture of virtual particle exchanges. The exchange must occur in a series of discrete steps, or skips, each carrying the same fluctuation energy and fluctuation momentum. The quantity ρ_K represents the density of potential skips.

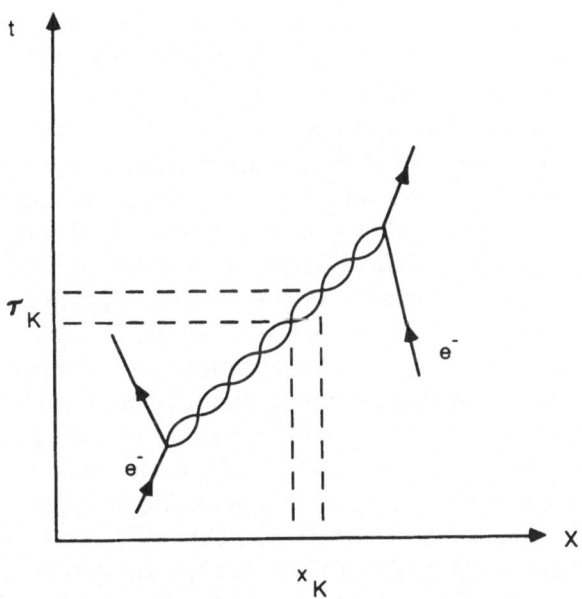

Figure 1. Skipping model of the interaction between two electrons. The closed loops represent propagating vacuum excitations. Later we will identify these excitations with transient electron-positron pairs. The probability of initiating or terminating a skipping process (i.e., of coupling charged positive energy particles to the electromagnetic field) should scale as the ratio of the Compton wavelengths of the transiently excited electron and positron ($\lambda_c = \lambda_e + \lambda_p \approx 2\hbar/m_e c$) to the average distance between an absorbing particle and an immediate neighboring vacuum particle ($x_K/2$). If the latter is identified with the radius of positronium in its lowest energy state $x_K/2 \approx 2\hbar^2/m_e e^2$, giving a ratio of $\lambda_c/(x_K/2) \approx e^2/\hbar c \approx 1/137$. The Compton wavelengths of the transiently excited electron and positron are summed since transfer of momentum via either of these particles may initiate or terminate a propagative skipping process.

The skipping process is pictured in Figure 1. The momentum carrier is represented by a series of closed loops, each representing a vacuum excitation whose energy extends over a fluctuation time τ_K and whose momentum extends over a fluctuation length x_K. The series is initiated by particles (to be called absorbers) which serve as sources or sinks of momentum. The exchange process is composed of repeated alternations between a completely localized aspect in which the energy and momentum fluctuations are quantized and a completely delocalized aspect, associated with the fact that the fluctuations disappear at the nodal points between skips.

The occurrence of skipping is a consequence of the requirement that macroscopic violations of conservation of momentum and conservation of energy should be forbidden. When a vacuum fluctuation results in a direct exchange of momentum with an absorbing particle energy conservation is transiently violated, but not momentum conservation. The momentum change of the absorbing particle is balanced by the momentum associated with the fluctuation, independent of the choice of coordinate system. The fluctuation must disappear within a time interval τ_K, otherwise an impermissible violation of energy conservation will occur. But the disappearance of the energy fluctuation is accompanied by the disappearance of the momentum. Since momentum has been conserved up to this point, its disappearance entails a violation of momentum conservation. This violation is restricted to a space interval x_K. Outside this interval the annihilated momentum must reappear, leading to a reappearance of the energy fluctuation. Since the energy fluctuation violates energy conservation, the process repeats until the momentum violation is annihilated by a second absorbing particle. Skipping is due to the fact that the quantization of energy and momentum fluctuations leads to a coupling of energy and momentum violations.

To formalize this argument we first consider the energy and momentum fluctuations which occur at each stage. For simplicity we consider only the x-component of momentum (the analysis is the same for the y and z components). Let $\Delta E(\tau_K, x_K)$ denote the energy fluctuation during a fluctuation interval and let $\Delta E(t_n, x_n)$ denote the energy fluctuation at the nodal points separating the fluctuation intervals. Similarly, let $\Delta p(\tau_K, x_K)$ denote the momentum fluctuation during a fluctuation interval and let $\Delta p(t_n, x_n)$ denote the momentum fluctuation at the nodal points between fluctuation intervals. According to Fig. 1, the fluctuation energies and momenta are zero at the instants between intervals and are equal to the values allowed by Eqs. (3) and

(20) during the intervals. Thus

$$\Delta E(\tau_K, x_K) \approx \frac{\hbar}{\tau_K} \qquad (\sigma t \approx \tau_K) \qquad (34\text{a})$$

$$\Delta E(t_n, x_n) \approx 0 \qquad (\sigma t_n \to \infty) \qquad (34\text{b})$$

$$\Delta p(\tau_K, x_K) \approx \frac{\hbar}{x_K} \qquad (\sigma x \approx x_K) \qquad (34\text{c})$$

$$\Delta p(t_n, x_n) \approx 0 \qquad (\sigma x_n \to \infty) \qquad (34\text{d})$$

According to Eqs. (34b) and (34d) no time or space points can be assigned to the fluctuation energy or momentum when these quantities equal zero. This is required by the uncertainty principle, but this requirement is consistent since the absence of a fluctuation cannot be assigned to a restricted set of spatial or temporal coordinates in any case. The delocalization of t_n through Eq. (34b) and of x_n through Eq. (34d) is compatible with the reasonable requirement that the fluctuation energy should disappear at the same points in space and time as the fluctuation momentum, and is compatible with the correlation of these points through the relation $x_n = ct_n$. Relations (34a-d) hold for all the intervals and nodes, including those directly exchanging momentum with absorbing particles.

We now use relations (34a-d) to write down the energy and momentum conservation violations which occur at each stage of the skipping process. Denote the energy conservation violation which occurs during a fluctuation interval by $\hat{E}(\tau_K, x_K)$ and the momentum conservation violation during the interval by $\hat{p}(\tau_K, x_K)$. Similarly denote the energy conservation violation at the points by $\hat{E}(t_n, x_n)$ and the momentum conservation violation at the points by $\hat{p}(t_n, x_n)$. Recalling that the energy conservation violations occur during the intervals and that the momentum conservation violations occur at the boundaries,

$$\hat{E}(\tau_K, x_K) \approx \frac{\hbar}{\tau_K} \qquad (\sigma t \approx \tau_K) \qquad (35\text{a})$$

$$\hat{E}(t_n, x_n) \approx 0 \qquad (\sigma t_n \to \infty) \qquad (35\text{b})$$

$$\hat{p}(\tau_K, x_K) \approx 0 \qquad (\sigma x \to \infty) \qquad (35\text{c})$$

$$\hat{p}(t_n, x_n) \approx \frac{\hbar}{x_K} \qquad (\sigma x_n \approx x_K) \qquad (35\text{d})$$

These relations hold for violations at all intervals and for all nodes, except for momentum violations at the nodes occurring between an energy violation and an absorbing particle. For these initiating and terminating nodes

$$\hat{p}(t_a, x_a) \approx 0 \qquad (\sigma x_a \to \infty) \qquad (35e)$$

where t_a and x_a indicate the time and space coordinates of an initiating and terminating node.

When the energy and momentum violations are zero they can occur anywhere in time or space. When they are nonzero they are restricted to the fluctuation intervals. Thus the momentum violation occurs at nodal points (other than initiating and terminating nodal points), but extends over the fluctuation interval (Eq. 35d). Relations (35a-d) differ from relations (34a-d) since nonexisting fluctuations of momentum may correspond to fluctuations away from conservation momentum.

The major feature of relations (35a-d) is that it is impossible to satisfy both conservation of energy and conservation of momentum except at initiating or terminating nodes. During the interval the momentum conservation violation is zero, but the energy conservation violation is $\hbar/\tau_K$. During the instants the energy conservation violation is zero, but the momentum conservation violation is $\hbar/x_K$. Thus once the skipping process is initiated by an absorbing particle there is no way of annihilating it except by another absorbing particle.

The following properties of the skipping process should be noted.

2.9.1. Applicability of uncertainty principle. Suppose that f skips occur between the initiation and termination of skipping. If t_1 is the instant of initiation and t_f the instant of termination (that is, $a = 1$ or f), we can write $t_1 - t_f = f\tau_K$. The average fluctuation energy during this interval is $\Delta E(\tau_K, x_K)$, due to the fact that the zero fluctuation values which occur at the nodal points only exist instantaneously (have negligible measure). Substituting into relation (34a)

$$\Delta E(\tau_K, x_K)(t_f - t_1) = f\Delta E(\tau_K, x_K)\tau_K \approx f\hbar \qquad (36)$$

Similarly,

$$\Delta p(\tau_K, x_K)(x_f - x_1) = f\Delta p(\tau_K, x_K)x_K \approx f\hbar \qquad (37)$$

where x_f and x_1 are spatial coordinates corresponding to t_f and t_1. The increase in uncertainty (to $f\hbar$ as compared to $\hbar$) is compatible with the applicability of the uncertainty principle to the exchange process as a whole and compatible with its applicability to each step ($f = 1$). The conservation laws are satisfied as closely as allowed by the uncertainty principle for the individual step. The increase in minimum uncertainty in Eqs. (36) and (37) results from the fact that it is impossible to satisfy both conservation of momentum and conservation of energy simultaneously. The uncertainty of the energy and momentum for the entire exchange are as small as possible compatible with these dual requirements. Relations (36) and (37) therefore express the fact that the values $\hbar/\tau_K$ and $\hbar/x_K$ must be maintained independent of the time and distance of the exchange. But this is compatible with the reversibility of the exchange since the uncertainty of the coordinates is independent of the number of skips.

2.9.2. Directionality. As t_n increases, x_n increases, that is, skipping proceeds from the initiating particle to the terminating particle without reversing itself. If reversal could occur the momentum violation $\hbar/x_K$ would double and would not be annihilated by a terminating particle (the initiating particle and the reversed skipping process would both have momentum $-\hbar/x_K$ rather than the allowable combination of $\hbar/x_K$ and $-\hbar/x_K$). But a propagating momentum violation can only arise as a consequence of the skipping process becoming separated from the initiating particle. If this were not the case each nodal point would act as an absorber and skipping could not occur.

2.9.3. Contiguity and propagation rule. The propagation rule for skipping can be written as

$$\hat{E}(t_{n+1} - t_n, x_{n+1} - x_n) \approx c\hat{p}(t_n, x_n) \tag{38}$$

where $t_{n+1} - t_n = \tau_K$ and $r_{n+1} - r_n = r_K$. That is, skipping propagates in a chain-like manner subject to the relations $x_K = c\tau_K$ and $x_n = ct_n$ despite delocalization at the nodal points. This is allowed by the fluctuation relations (34a-d), but not required by them. However, it is required by the violation relations (35a-d) since the momentum violation can only extend over an interval x_K. Thus the fact that absence of fluctuation or the absence of violation can occur anywhere in space does not imply that the fluctuation violations can disappear or appear anywhere in space (except that the initiating and terminating fluctuation can of course occur in the neighborhood of any absorbing particle).

2.9.4. Momentum correlations and entropy conservation. The microscopic reversibility of the skipping process is connected to the fact that each skip has the same momentum. The propagation rule (Eq. 38) implies that the skip occurring during the time interval $t_{n+1} - t_n$ occurs in a spherical shell of outer radius r_{n+1} and inner radius r_n. The skip is confined to one fluctuation volume, given by $4\pi(r_n - r_{n-1})^3/3$, but the number of eligible fluctuation volumes increases as $4\pi(r_n^3 - r_{n-1}^3)/3$. The probability of the skip occurring in any of these eligible regions thus appears to decrease with n, in violation of microscopic reversibility. However, this is only true if the information about the direction of the initiating particle is discarded. The skip which occurs during the interval $t_{n+1} - t_n$ always occurs in the unit fluctuation volume whose coordinate values are compatible with the requirement that the momentum of each skip must be opposite to that of the initiating particle. Thus the uncertainty $\hbar/(x_{n+1} - x_n)$ actually does not depend on n, but instead always remains equal to the uncertainty of the direction of the initial skip.

It is useful to compare these features to those present in quantum field theories. In the classical picture a massless virtual momentum carrier (such as a virtual photon) can travel a distance $\hbar c/2\Delta E \approx c/2\nu = \lambda/2$, where the energy ΔE can assume a range of values. The half-wavelength restriction prevents virtual quanta from transmitting information, which is necessary for consistency with relativity since the time-order of emission and absorption cannot be distinguished. In the skipping model the energy can only assume one value, depending on the fluctuation density. Skips cannot return to their source, though they remain correlated to it. This leads to no violation of the uncertainty principle because every excitation is annihilated in time τ_K. Since skips always propagate away from their source, there is no self-interaction, but the absence of a self-interaction energy is not detectible since it is impossible to distinguish whether the excitation absorbed by the source originates in the immediate surounding vacuum (in which case it emits) or whether it originates from the neighborhood of another absorbing particle. The finiteness of the self-energy is due to the quantization of momentum transferred. The skipping process cannot carry information since the pattern of skipping is an invariant feature determined by the fluctuation density.

2.10. Zero rest mass property. The constancy of skipping velocity in all inertial systems in the absence of mass and charge is a thermodynamic requirement, as demonstrated in Section 2.5. The relation $r_K^2 = c^2\tau_K^2$ is justified by this argument. Since $\Delta E = m_0 c^2/(1 - v^2/c^2)^{\frac{1}{2}} \approx \hbar/\tau_K$ despite the fact that $v = c$, the skipping process must have the zero rest mass property. As shown in Section 2.5, the relation between skipping energy and momentum, $\Delta E = \Delta pc$, may be obtained by equating relations (34a) and (34c) and utilizing the thermodynamic requirement $x_K = c\tau_K$. The zero rest mass property of skipping is also implied by this relation taken together with the relativistic energy momentum relation.

To analyze the rest mass properties of skipping more precisely, we first observe that the energy and momentum of skipping are purely fluctuation quantities. Thus the relativistic energy momentum relation

$$E = \sqrt{p^2 c^2 + m_0^2 c^4} \tag{39}$$

may be written as

$$\Delta E = \sqrt{\Delta p^2 c^2 + \Delta m_0^2 c^4} \tag{40}$$

where ΔE and Δp are properly interpreted as σE and σp. Since m_0 and c are not usually regarded as quantitities which fluctuate, Eq. (41) reduces to $\sigma = \sigma pc$, corresponding to the relation $\Delta E = \Delta pc$. Thus the zero rest mass property of skipping may be viewed as simply expressing the fact that m_0 does not fluctuate. But note that this does not preclude the possibility that particles with nonzero rest mass could provide a substrate for skipping (since Eq. (39) would apply to such a particle, not Eq. (40)).

2.10.1. Interval and nodal point rest mass. The dependence of the zero rest mass property on light velocity is not the same during the intervals and at the nodal points. This is due to the fact that ΔE and Δp are zero at the nodal points. To make matters more precise the relation between energy and momentum during the skip may be written as

$$\Delta E(\tau_K, x_K)^2 = \Delta p(\tau_K, x_K)^2 c^2 + m_K(\tau_K, x_K)^2 c^4 \tag{41}$$

where $m_K(\tau_K, x_K)$ is the rest mass formally assigned to the skipping process during the interval (τ_K, x_K). Substituting the fluctuation relations (34a) and (34c)

$$m_K(\tau_K, x_K) = 0 \tag{42}$$

where it is assumed that $x_K = c\tau_K$. At the points between the skips

$$\Delta E(t_n, x_n)^2 = \Delta p(t_n, x_n)^2 c^2 + m_K(t_n, x_n)^2 c^4 \tag{43}$$

where $m_K(t_n, x_n)$ is the mass formally assigned to the skipping process at these points. Substituting the fluctuation relations (34b) and (34d)

$$m_K(t_n, x_n) = 0 \qquad (\sigma t_n \to \infty, \sigma x_n \to \infty) \tag{44}$$

Thus the rest mass formally associated with the skipping process is zero during the intervals if c is taken as the skipping velocity and the rest mass formally associated with skipping at the nodal points is zero independent of the velocity, consistent with the delocalization of the process at these points.

2.10.2. Violation masses. The energy-momentum relation may also be employed to assign a formal rest mass correspondence to the violations of energy and momentum conservation. For the intervals

$$\hat{E}(\tau_K, x_K)^2 = \hat{p}(\tau_K, x_K)^2 c^2 + \hat{m}_0(\tau_K, x_K)^2 c^4 \tag{45}$$

where $\hat{m}_0(\tau_K, x_K)$ is the violation rest mass formally assigned to the skipping process during the interval (τ_K, x_K). Substituting the violation relations (35a) and (35c)

$$\hat{m}_0(\tau_K, x_K) \approx \pm \frac{\hbar}{x_K c} \tag{46}$$

with $\sigma t = \tau_K$ and $\sigma x \to \infty$.

At the instants

$$\hat{E}(t_n, x_n) = \hat{p}(t_n, x_n)^2 c^2 + \hat{m}_0(t_n, x_n)^2 c^4 \tag{47}$$

where $\hat{m}_0(t_n, x_n)$ is the rest mass formally assigned to the skipping process at the nodal points (other than the absorbing points). Substituting the violation relations (35b) and (35d)

$$\hat{m}_0(t_n, x_n) \approx \pm \frac{i\hbar}{x_K c} \tag{48}$$

with $\sigma x_n \sim x_K$ and $\sigma t_n \to \infty$. At the absorbing points relation (35e) replaces (35d), giving

$$\hat{m}_0(t_a, x_a) \approx 0 \tag{49}$$

with $\sigma x_a \to \infty$ and $\sigma t_a \to \infty$.

2.10.3. Applicability of energy-momentum relation. The point of view taken in noncovariant perturbation theory is that momentum conservation holds during the exchange process, but that energy conservation fails. In the skipping model momentum conservation holds for the entire exchange process between two absorbers, but fails at the endpoint of each constituent skip. The point of view usually taken in covariant perturbation theory is that energy and momentum are both formally conserved, but the relativistic energy-momentum relation fails during the process. This is due to the fact that energy and momentum are treated on the same footing and it therefore appears unnatural to regard energy as fluctuating in a different way than momentum. In the skipping model energy and momentum both fluctuate and both violate conservation laws during the exchange, the only difference being that momentum is conserved at the points of absorption. As a consequence, the energy-momentum relation may be viewed as holding while at the same time treating energy and momentum in a uniform manner so far as fluctuation and conservation law violations are concerned.

2.11. Bosons as skipping processes. A real photon can also be interpreted as a skipping process. In this case each of the skips carries macroscopically observable energy ($E = h\nu$) and momentum ($p = E/c = h/\lambda$) which arise from the acceleration of an absorbing particle (apart from the acceleration due to the virtual exchange itself). In the absence of such preexisting acceleration the momentum change of the absorber is due solely to fluctuation and is balanced by the momentum carried by an initial or terminal skipping event. When the absorber is accelerating this is not possible unless the initial skipping event also carries away the momentum arising from the acceleration. In continuous field theories momentum can flow away from the absorber because the half-wavelength restriction on virtual photons does not apply to real photons. In the skipping model the momentum can flow away from the absorber because the acceleration of the absorber decorrelates its momentum from the fluctuation momentum of the skipping process. As a consequence real photons and other real bosons cannot mediate an inverse square interaction.

According to the skipping model the energy of a photon can be written as $E = \epsilon + \Delta E$, where ϵ is the real (conservation law obeying) energy and ΔE is the virtual energy annihilated and created at each skip. To be consistent with the relation $E = h\nu$ for photons, ΔE must be macroscopically unobservable, as virtual energies should be. Similarly the momentum of the photon can be written as $p = P + \Delta p$, where P is the real (conservation law obeying) momentum and Δp is the fluctuation momentum annihilated and created at each skip. To be consistent with the relation $p = h/\lambda$, the fluctuation momentum should also be macroscopically unobservable. Later (Section 3.7) we will show that if ρ_K is inhomogeneous the skipping energy can remain constant only if each skip is associated with a local expansion or contraction of the vacuum density. This contraction and expansion carries the energy of real bosons and corresponds to the wave aspect of radiation.

Skipping cannot be directly observable since any observation would allow the skipping momentum to be transferred to the measuring apparatus, thereby terminating the process. Since the energy violation and momentum violation are linked, any observation which eliminates one eliminates the other. If r_K is small the conservation law violations are high, but the probability of interacting with the measuring apparatus is low. If r_K is large the probability that the skipping process will interact with the measuring apparatus is high, but the conservation law violations are small. In the first instance the contributions of ΔE and Δp to E and p are large, but the conditions for measurement do not exist. In the second instance, where the conditions for measurement are met, the contributions of ΔE and Δp are negligible.

2.13. Review of Section 2 assumptions. The derivation of the inverse square law is based on three assumptions. The first is that the time-energy uncertainty principle is applicable to virtual particle exchanges mediating a force. The second is that the transition probability of first order, time independent perturbation theory is the appropriate one. The third assumption is that the relation between the fluctuation energy required to achieve any given momentum transfer is linearly dependent on distance. Initially the first order perturbation formula may appear to be a lowest order approximation rather than a bona fide physical assumption. But better approximations do not work, a feature which is rationalized by the necessity of introducing fluctuation quantities. If the inverse square law is taken as an assumption, the logic can be turned around and the linearity assumption moved to the status of a deduction. The linear relation between energy and distance follows immediately by equating the denominator of the perturbation formula (Eq. 1) with the denominator of the inverse square law (multiplied by a dimensionally appropriate factor). The time-energy uncertainty principle sets $\Delta E \tau / \hbar \approx 1$ in the numerator of Eq. (1) since the energy shift is identified with the energy fluctuation conjugate to the fluctuation time.

The occurrence of the fluctuation time in the inverse square law is a consequence of the fact that this identification and the linearity relation taken together severely restrict the amount of momentum that can actually be transferred in a virtual process (cf. Eq. 8). This result is significant in that the occurrence of divergences in quantum field theories is due to the absence of such inbuilt cutoffs. The introduction of fluctuation length and fluctuation number (Eq. 10) is also a consequence of the assumption that the constancy of light velocity is valid at this scale of phenomena. However, the assumption of constant light velocity can also be moved to the status of a deduction by utilizing the position-momentum uncertainty principle and forbidding violations of macroscopic thermodynamics. This result is significant insofar as it suggests that the necessity of four-space descriptions in the large may be compatible with the validity of the three-space descriptions in the small, which form the starting point of the present theory. The construction of electromagnetic force laws for moving charges requires the principle of relativity to be added to the list of principles used. Force laws with an apparent noninverse square character may be obtained by classifying the specializing assumptions which are possible in the fluctuation density framework.

3. Vacuum Particle Interpretation

3.1. Survey of interpretations. Force laws expressed in terms of fluctuation quantities (such as Eqs. 16 and 18) imply the occurrence of transient excitations or skips (represented by the closed loops of Fig. 1). But the mathematical form of these force laws is not by itself sufficient to determine the physical nature of the excitations. To fully demonstrate that the skipping model is tenable, it is necessary to show that at least one internally consistent physical interpretation exists. It is useful to begin by classifying the most obvious possibilities, and to indicate the reasons for choosing the one that identifies the skipping process with a chain of transient particle-antiparticle pairs.

    i. *Intrinsic space-time interpretation.* In this interpretation the fluctuation lengths and fluctuation times which define the closed loops are intrinsic properties of space-time, such as intrinsic geometrical properties. The objection to this interpretation is that the fluctuation times and fluctuation lengths are different for the electromagnetic and gravitational fields, therefore two different intrinsic properties would have to cohabit the same space-time.

    ii. *Particle property interpretation.* In this interpretation fluctuation lengths and fluctuation times are intrinsic properties of the virtual momentum carrier. The objection to this interpretation is that mass and charge alter fluctuation times and lengths (Eqs. 17 and 19), leading to the unreasonable need for an additional force mechanism which enables the intrinsic properties of the momentum carrier to be controlled by the distribution of mass and charge in the universe.

iii. *Vacuum particle interpretation.* This identifies the closed loops with excitations occurring in a Dirac sea of negative energy particles. To demonstrate the physical tenability of the vacuum particle interpretation it is necessary to show that it satisfies all the properties of the skipping process and that the altering effects of mass and charge on vacuum density can be attributed to the gravitational and electromagnetic interactions. Two cases should be considered:

a. *With pair production:* In this case the closed loops are identified with short-lived particle-antiparticle pairs, with the antiparticle interpreted as a hole in a Dirac sea of negative energy particles. The objection which must be overcome is that the fluctuation energy required for pair creation increases with vacuum density, whereas the mass defined by the gravitational interaction of these particles decreases with vacuum density (Eq. 17).

b. *Without pair production:* In this case the closed loops are identified with vacuum particle excitations, but the fluctuation energies required are allowed to be smaller than that required for pair production. The excitation may involve an energy shift of a vacuum point particle or it may involve an internal excitation of a compound vacuum particle. The objection is that the availability of such low energy states of excitation would initiate a collapse of observable particles into these states.

An additional consideration is that the vacuum particle interpretation provides a natural explanation for the fact that the gravitational force is weaker than the electromagnetic force. The fluctuation density ρ_K represents the density of skipping excitations. According to the vacuum particle interpretation, ρ_K must be less than or equal to the density of vacuum particles which can contribute to particle-antiparticle creation. All known particles couple through mass, but only some couple through charge. Assuming that this holds true for vacuum particles, ρ_{grav} must be higher than ρ_{em}. But according to Eqs. (16) and (18), $\rho_{grav} > \rho_{em}$ implies that

$$\frac{F_{grav}}{F_{Coul}} \approx \frac{\hat{A}_{grav}\rho_{em}^{\frac{4}{3}}}{\hat{A}_{em}\rho_{grav}^{\frac{4}{3}}} < 1 \qquad (50)$$

The remainder of this section will focus on the construction of a vacuum particle interpretation. The first step (Section 3.3) is to show that vacuum particle excitations have the same propagation and rest mass properties as the skipping process. The second step (Sections 3.5 and 3.6) is to show that the objection to the pair production mechanism (iii-a) can be overcome by admitting that the velocity of light is altered in the presence of mass (as in general relativity). Any difference between gravitational mass and fluctuation energy can also be absorbed in the excess kinetic energy of the excited pairs, a feature important for vacuum particles mediating the electromagnetic field. A significant result is that the alterations of vacuum densities produced by mass and charge require the skipping process to be accompanied by vacuum density waves reminiscent of electromagnetic and gravitational waves (Section 3.7). Another conceptually significant feature is that the vacuum particle interpretation leads directly to Mach's principle and the principle of equivalence (Section 3.8). The relations between vacuum particle mass and vacuum density suggests that sharply varying vacuum densities are a potential source of pulsar-like effects (Section 3.9).

In presenting the demonstration of physical tenability we use the Dirac picture of the vacuum as the appropriate physical background. One reason is that it is convenient to demonstrate the physical tenability of the skipping model from the standpoint of particle interactions. An interesting question is whether it is possible to interpret the antiparticles in transient pair formation in the fashion of Feynman as positive energy particles travelling backward in time rather than as a hole in a sea of negative energy particles. This is possible as long as it is admitted that the presence of mass and charge can alter the potentialities for pair production (by altering the wave function of the vacuum), just as they alter the structure of the Dirac sea.

3.2. Vacuum particle skipping and boson structures. The virtual violation of energy required for observable pair creation is $\Delta E \geq 2m_0 c^2$, where m_0 is the rest mass of either the particle or antiparticle. This is therefore the energy violation required for each step in a skipping process if $\Delta E/2$ is identified with the rest mass energy of the vacuum particles which provide the substrate for skipping. But in free space pair creation or annihilation violates conservation of momentum. This is due to the fact that the total momentum of the particle and antiparticle is zero in a coordinate system in which their center of mass is at rest, but cannot be zero in an arbitrary coordinate system. Pair production or annihilation can occur without momentum violation only when it is coupled to an absorbing particle since in this case the momentum change of the pair resulting from a transformation to a new coordinate system is always compensated by the momentum change of the absorber.

Transient excitations of vacuum particles thus fulfill exactly the condition required for skipping. A virtual pair creation process occurring in the neighborhood of an absorbing particle violates conservation of energy, but not conservation of momentum. Conservation of energy thus requires the pair to annihilate in a short amount of time, but this creates a momentum violation which can only extend over a short interval of space. The momentum violation annihilates itself by creating a second pair, and so forth, until the momentum violation is annihilated by an absorbing particle. But it should be noted that this argument would hold for any vacuum particle excitation, whether or not it involved pair production, since the momentum of the excited particle would depend on the choice of coordinate system. The only difference is that the condition $\Delta E \geq 2m_0c^2$ would not apply. The identification of the closed loops in Fig. 1 with pair production cannot be deduced from the coupling of energy and momentum violations in skipping, but rather is dictated by the assumption that the Dirac vacuum is the appropriate physical background.

The world line of a real boson may also be interpreted as a propagating chain of short-lived particle-hole formations. As in the case of virtual bosons, skipping is due to conservation law violations connected with initial pair production and with pair creation and annihilation processes which are uncoupled to the momentum change of an absorbing particle. The difference is that the transient excitations of vacuum fermions corresponding to real bosons carry extra momentum, due to the fact that they arise in conjunction with accelerations of absorbing particles. In Section 3.7 we will show that skipping processes must in general be accompanied by a compression-expansion wave in the vacuum density and that the real energy and momentum of a real boson is carried by this wave.

The transient particle-antiparticle pair may be either in a singlet or a triplet state. If the particle and antiparticle are spin 1/2 particles the pair has spin 0 in the antiparallel configuration and spin 1 in the parallel configuration. Since the photon has spin 1, the transient electron and positron which constitute it should be in the parallel (triplet state) configuration. Individually these constituent particles obey the exclusion principle, but since the composite system has an integral spin property it obeys Bose-Einstein statistics. From spin 3/2 fermions it is possible to construct spin 3 bosons, and so forth for all odd spin bosons. To construct spin 2 bosons (e.g., spin 2 gravitons) it is necessary to assume that the pair itself carries angular momentum, presumably due to the structure of the density wave which accompanies it, or that the structures formed are more complicated than pairs. The formation of a transient pair is always accompanied by a transient violation of conservation of angular momentum, with the transient appearance of angular momentum $\hbar/2$ in the case of spin 1 skipping processes and the transient disappearance of angular momentum $\hbar/2$ in the case of spin 0 skipping processes. The propagation of these angular momentum violations is coupled to the propagation of the energy and momentum violations.

By constructing bosons in this way it is possible to interpret them in the framework of a hole theory. However, antibosons must be interpreted as skipping processes propagating backwards in time. The identification of the antiparticles occurring in skipping processes with holes therefore does not exclude the same type of antiparticle from occurring as a consequence of reversal of time in a different type of process.

3.3. Annihilation of vacuum particle rest mass. To express the vacuum particle rest mass in terms of fluctuation quantities first note that in virtual pair creation the total relativistic energy of the negative energy particle is converted to the total relativistic energy of the positive energy particles. The fluctuation energy may be written as $\Delta E = E_+ - E_-$, where $E_+(> 0)$ denotes the total relativistic energy of either the positive energy particle or positive energy hole and $E_-(< 0)$ denotes the total relativistic energy of the negative energy particle. In the absence of a potential E_+ and E_- may be written as

$$E_+ = T_+ + m_0 c^2 \tag{51a}$$
$$E_- = T_- - m_0 c^2 \tag{51b}$$

where $T_+(> 0)$ is the kinetic energy of the vacuum particle when excited to a positive energy state and $T_-(< 0)$ is its kinetic energy in the negative energy state. Thus,

$$\Delta E = T_+ - T_- + 2m_0 c^2 \tag{52}$$

But $\Delta E \approx \hbar/\tau_K$, giving

$$m_0 \approx \frac{\hbar}{2\tau_K c^2} + \frac{T_- - T_+}{2c^2} \tag{53}$$

Expressed in terms of fluctuation lengths

$$m_0 \approx \frac{\hbar}{2r_K c} + \frac{T_- - T_+}{2c^2} \tag{54}$$

and in terms of fluctuation densities

$$m_0 \approx \frac{\hbar \rho_K^{\frac{1}{3}}}{2c} + \frac{T_- - T_+}{2c^2} \tag{55}$$

These equations may be simplified by noting that the kinetic energy of the vacuum particles excited throughout a skipping process must all be the same. If this were not the case the fluctuation energy and fluctuation momentum would vary from step to step. If $T_- < 0$ a negative energy particle with $T_- = 0$ could always occupy the hole left by the excited vacuum particle, thereby leading to a change in the fluctuation energy and momentum during the skipping process. We can therefore assume that the initial fluctuation either occurs from the ground state $T_- = 0$ or that if it does not it is coupled to a transition which in effect sets $T_- = 0$.

The vacuum particle interpretation is tenable only if the nonzero rest mass defined by Eq. (55) can be reconciled with the zero rest mass property of the skipping process. It is possible to meet this requirement only if the rest mass of a vacuum particle excited in skipping is converted to pure kinetic energy. To prove this we express the relation $\Delta E = E_+ - E_-$ in terms of the allowed value of ΔE and the total relativistic energy of E_+ and E_-, yielding

$$\frac{\hbar}{\tau_K} \approx \frac{m_0(\tau_K, x_K)c^2}{\sqrt{1 - v_+^2/c^2}} + \frac{m_0(t_n, x_n)c^2}{\sqrt{1 - v_-^2/c^2}} \tag{56}$$

Here v_+ denotes the velocity of the particle in its positive energy state and v_- denotes its velocity in the negative energy state. During skipping the rest mass in the positive energy state (denoted by $m_0(\tau_K, x_K)$ to indicate that it is the rest mass during the skipping interval) should be the same as the rest mass in the negative energy state (denoted by $m_0(t_n, x_n)$ to indicate that it is the rest mass at a nodal point). However, as the rest mass is the property under investigation it is convenient to distinguish the interval and nodal point rest mass and to treat their relation as unknown.

The fluctuation momentum may be expressed as

$$\Delta p(\tau_K, x_K) \approx \frac{\hbar}{x_K} = \frac{\hbar}{\tau_K c} \approx \frac{m_0(\tau_K, x_K)c}{\sqrt{1 - v_+^2/c^2}} + \frac{m_0(t_n, x_n)c}{\sqrt{1 - v_-^2/c^2}} \tag{57}$$

where we utilize the already established relation $r_K = c\tau_K$. Since the rest mass m_0 is independent of the choice of the negative energy state, we can choose the highest negative energy state without any loss of generality. In this state $p = 0$, therefore $v_- = 0$. Thus Eq. (57) becomes

$$\Delta p(\tau_K, x_K) \approx \frac{m_0(\tau_K, x_K)c}{\sqrt{1 - v_+^2/c^2}} + m_0(t_n, x_n)c \qquad (58)$$

But by definition $\mathbf{p} = m(v)\mathbf{v}$. Thus

$$\Delta p(\tau_K, x_K) = \frac{m_0(\tau_K, x_K)v_+}{\sqrt{1 - v_+^2/c^2}} \qquad (59)$$

where $\Delta p(\tau_K, x_K)$ and v_+ are corresponding components of momentum and velocity and we again take the case of $v_- = 0$. Equating (59) and (60), we find

$$v_+ \approx c + \frac{\sqrt{1 - v_+^2/c^2}\, m_0(t_n, x_n)c}{m_0(\tau_K, x_K)} \qquad (60)$$

We can express $m_0(\tau_K, x_K)$ in terms of $m_0(t_n, x_n)$ in this equation by using the relation $\Delta E = E_+ - E_-$ to write (for $v_- = 0$)

$$\frac{\hbar}{\tau_K} \approx E_+ + m_0(\tau_n, x_n)c^2 \qquad (61)$$

But E_+ may also be written as

$$E_+ = \frac{m_0(\tau_K, x_K)c^2}{\sqrt{1 - v_+^2/c^2}} \qquad (62)$$

Eliminating E_+ from Eq. (61) and (62) and solving for $m_0(\tau_K, x_K)$

$$m_0(\tau_K, x_K) \approx \frac{\left[\hbar/\tau_K - m_0(t_n, x_n)c^2\right]\sqrt{1 - v_+^2/c^2}}{c^2} \qquad (63)$$

Substituting into Eq. (60)

$$v_+ \approx c + \left\{ \frac{m_0(t_n, x_n)c^3}{\hbar/\tau_K - m_0(t_n, x_n)c^2} \right\} \qquad (64)$$

But the relation $\Delta E \geq 2m_0(t_n, x_n)c^2$ implies $\hbar/\tau_K > m_0(t_n, x_n)c^2$. Thus the term in curly brackets in Eq. (64) is positive if it is assumed that $m_0(t_n, x_n) > 0$. However, this leads to the unacceptable consequence that $v_+ > c$. The assumption of positive $m_0(t_n, x_n)$ must therefore be rejected, giving $m_0(t_n, x_n) = 0$, which in turn implies $v_+ = c$.

Again eliminating E_+ from Eqs. (61) and (62), but this time solving for $m_0(t_n, x_n)$,

$$m_0(t_n, x_n) \approx \frac{\hbar}{\tau_K c^2} - \frac{m_0(\tau_K, x_K)}{\sqrt{1 - v_+^2/c^2}} \tag{65}$$

Substituting in Eq. (60)

$$v_+[\tau_K c m_0(\tau_K, x_K)] \approx \hbar \sqrt{1 - v_+^2/c^2} \tag{66}$$

But since $v_+ \approx c, m_0(\tau_K, x_K) \approx 0$. Thus

$$m_0(\tau_K, x_K) \approx m_0(t_n, x_n) = 0 \tag{67}$$

Substituting either $m_0(t_n, x_n)$ or $m_0(\tau_K, x_K)$ into Eq. (51a) gives $E_+ \approx T_+$. That is, all the rest mass is converted into pure kinetic energy when the vacuum particles are recruited for a skipping process. Recalling that $E_+ = (p^2 c^2 + m_0 c^4)^{\frac{1}{2}}$, we can also write $E_+ \approx pc$, subject to the condition $E_+ \approx T_+ \geq 2m_0 c^2$ (since E_+ now equals ΔE for the case $v_- = 0$). Thus the relation between energy and momentum of vacuum particles recruited for skipping corresponds to the energy-momentum relation for massless bosons.

The equality of $m_0(t_n, x_n)$ and $m_0(\tau_K, x_K)$ is consistent with the reasonable requirement that vacuum particles excited in a skipping process have the same rest mass both during the intervals and at the nodal points. The zero value of these masses merely expresses the fact that the energy of a particle traveling with velocity c must be purely kinetic. The rest masses $m_0(t_n, x_n)$ and $m_0(\tau_K, x_K)$ may therefore be identified with the rest masses $m_K(t_n, x_n)$ and $m_K(\tau_K, x_K)$ formally assigned to the skipping process (Eqs 41 and 43). The important difference is that the vacuum particle description starts with negative energy particles which have a nonzero rest mass as ascertained by the amount of energy required to transfer these particles to the positive energy state (see Eq. 52).

The argument presented may be viewed as convincing in that it extracts a zero rest mass boson property from nonzero rest mass fermions.

However, it might also be argued that the zero rest mass property of vacuum particles both during the intervals and at the nodal points invalidates the initial premise that vacuum particles with nonzero rest mass provide the substrate for skipping. To complete the argument it is therefore necessary to show that $m_0(t_n, x_n) = m_0(\tau_K, x_K) = 0$ does not contradict $m_0 > 0$ in Eq. (54). To do this we employ the violation masses formally assigned to the energy and momentum conservation violations which occur in skipping (cf. Eqs. 46 and 48). The idea is that the conversion of vacuum particle rest mass to pure kinetic energy in skipping (that is, the conversion from a nonzero to a zero rest mass) is a transient violation of conservation laws allowed by the uncertainty principle. As such it can lead to no observable mass, otherwise the conservation law violations would become macroscopically observable.

To verify this claim first write Eq. (52) as

$$\frac{\hbar}{\tau_K} \approx 2m_0^2 + T_+ - T_- \tag{68}$$

Since $\Delta E \approx \hbar/\tau_K$ is fixed and T_- may be taken as zero, T_+ must increase to compensate the annihilation of rest mass when m_0 is replaced by $m_0(t_n, x_n) = 0$ or $m_0(\tau_K, x_K) = 0$. Thus the relation $m_0 > 0$ for vacuum particles as ascertained by the energy requirement of skipping and the relation $m_0 \approx 0$ as ascertained by the velocity requirement are both consistent with conservation of energy.

Substituting the violation mass relation (Eqs. 46 and 48) into Eq. (68)

$$\hat{m}_0(\tau_K, x_K) \approx \pm \left(2m_0 + \frac{T_+ - T_-}{c^2} \right) \tag{69a}$$

and

$$\hat{m}_0(t_n, x_n) \approx \pm i \left(2m_0 + \frac{T_+ - T_-}{c^2} \right) \tag{69b}$$

But during skipping, m_0 is properly replaced by $m_0(\tau_K, x_K)$ or by $m_0(t_n, x_n)$, giving

$$\hat{m}_0(\tau_K, x_K)c^2 \approx \pm (T_+ - T_-) \tag{70a}$$

and

$$\hat{m}_0(t_n, x_n)c^2 \approx \pm i(T_+ - T_-) \tag{70b}$$

where the values of $\hat{m}_0(\tau_K, x_K)$ and $\hat{m}_0(t_n, x_n)$ are fixed by τ_K. The annihilation of m_0 by skipping thus exists by virtue of a violation of

conservation of energy. During the intervals the violation kinetic energy
is real. At the nodal points it is pure imaginary, corresponding to
the fact that it is the momentum violation, not the energy violation,
which annihilates the vacuum particle rest mass at these points. The
coexistence of the nonzero and zero rest mass properties of vacuum
particles is of the same nature as the coexistence of the energy and
momentum conservation laws with the violations of these laws allowed
by the uncertainty principles.

The argument does not apply at the absorbing points, since at
these places $\hat{m}_0(t_a, x_a) = 0$. Thus the rest mass of vacuum particles
initiating or terminating skipping processes by transferring their momentum to absorbing particles cannot be zero. This corresponds to the
fact that the fluctuations of these particles do not violate conservation
of momentum and therefore the assumption $v_+ = c$ is not a thermodynamic requirement. This exception is in agreement with the occurrence
of massive virtual photons in the neighborhood of electrons in quantum
electrodynamics.

3.4. Trapped skipping and origin of gravitational mass. The analysis of vacuum particle skipping has been based on the assumption
that vacuum particles in effect have gravitational mass (otherwise they
could not have negative energy). The violation mass argument estab-
lishes that no formal contradiction ensues from building zero rest mass
bosons from such particles. To be fully consistent, however, the vacuum
particle interpretation requires the mass of "negative energy" particles
to be interpreted as a latent property inherited from the positive energy form of the particle. The difficulty which otherwise arises is that
vacuum particles would have gravitational mass only if they exchange
momentum through skipping processes, that is, if they are absorbing
particles. But they cannot at the same time serve as substrate particles
and absorbing particles. Insofar as they are absorbing particles they
would terminate any skipping processes which excite them.

Vacuum particles must have two modes of existence to allow them
to meet these dual requirements. The first is as potential substrate for
skipping processes. The second is as a skipping process trapped in a
fluctuation volume. Such trapping occurs because a vacuum particle
undergoing an improper fluctuation ($\Delta E > \hbar/\tau_K$ and $\Delta p > \hbar/x_K$) can
serve as a transient absorber. Ordinarily such an improper transient
pair would fluctuate out of existence without producing any new energy
or momentum violations. However, suppose that a second transient
pair transfers some momentum to this first pair and that at the end

of the transfer one of the pairs has a proper fluctuation ($\Delta E \approx \hbar/\tau_K$ and $\Delta p \approx \hbar/x_K$) and one an improper fluctuation. In the simplest case a vacuum particle undergoing a proper fluctuation transfers its momentum to the improperly fluctuating vacuum particle. The decay of the properly fluctuating pair produces the momentum conservation violation, thereby initiating a propagative skipping process. The decay of the improperly fluctuating particle is unusual, however, since it leads to the annihilation of the fluctuation energy and momentum absorbed from the properly fluctuating particle, as well as to the annihilation of its own fluctuation energy and momentum. Thus its decay now leads to a conservation of momentum violation (its absorption of momentum from the properly fluctuating vacuum particle did not violate conservation of momentum). Since the energy violation is greater than $\hbar/\tau_K$ and the momentum violation is greater than $\hbar/x_K$ the decay of the improperly fluctuating particle must occur in a time less than the fluctuation time and must be confined to a volume smaller than the fluctuation volume. The annihilation of the improper pair in one subregion of the fluctuation volume must thus be recoupled to its creation in another subregion, leading to a trapped skipping process which can only be terminated by a momentum exchange with a propagative skipping process.

In this way a vacuum particle obtains gravitational mass through its existence as a skipping-in-place absorber. Since it can only absorb when it undergoes improper fluctuations this gravitational mass must be regarded as an average property. When it is not undergoing an improper fluctuation a vacuum particle serves as a substrate particle for propagative skipping. Skipping in place does not contradict the thermodynamic argument that skipping must always propagate with velocity c since this type of skipping does not mediate forces. Any observation of skipping in place would entail a momentum transfer which would serve to terminate skipping.

The origin of gravitational mass can be thought of as a bootstrapping process. Vacuum particles serve as a substrate for skipping, which allows them to exert forces on each other when they are undergoing trapped skipping. The gravitational mass resulting from these forces alters the energy required for skipping, thereby altering the forces exerted, and so on, until self-consistency (or stability) is reached. When the vacuum particles are not skipping in place no forces act on them, but their spatial distribution is controlled by the forces which act on them when they are in this mode. This spatial distribution in turn controls the virtual energy required for skipping. This is an integral part of the bootstrap process. An important point to note is that the vacuum density is in one respect more controlling in this bootstrap process than is the gravitational mass. As soon as the gravitational mass exceeds the fluctuation energy required for skipping, the force mediated by this skipping process is turned off altogether.

The trapped skipping mechanism is also required for a consistent interpretation of the electromagnetic interaction. Since vacuum particles can absorb only when they are skipping in place they can evince electric charge only when they are in a positive energy state. Thus vacuum particles interact with each other through skipping processes only in the form of positive energy particles, whether the interaction is gravitational or electromagnetic. But there is a significant difference between these two interactions.

When a charged vacuum particle is promoted two positive energy masses are produced, whereas positive and negative charges are produced. The simplest interpretation is that electromagnetic vacuum particles possess negative charge in their negative energy state, but cannot express the charge since they cannot serve as absorbing particles. When a charged vacuum particle is promoted its negative charge becomes manifest. This cannot be the case for the antiparticle, since this is a vacuum hole from which charge is absent. All interactions between charged particles in either a transient or permanent positive energy state are repulsive, but this repulsion leads to an apparent attractive interaction between a particle and hole. This is not possible for the gravitational interaction since the hole represents positive mass as much as the particle does. Thus the skipping-in-place mechanism is consistent with the fact that the gravitational interaction is always attractive even though the Dirac equation has both positive and negative frequency solutions, whereas the sign of the interaction between charges depends on the product of the signs of the charges.

Later (in Section 3.8) we will consider how the interactions between skipping-in-place particles and between such particles and permanent positive energy particles determines the structure of the vacuum density. Mass will eventually be associated with depressions in the vacuum density. Particles and antiparticles will be seen to have positive mass because they are both associated with such depressions. The negative energy of vacuum particles may be interpreted as a negative frequency phenomenon; but it is also possible to view negative energy (or mass) as a property formally assumed by vacuum particles due to the structure of the vacuum density. The infinite but self-cancelling charge density formally attributed to the Dirac vacuum is eliminated by the requirement that all direct interactions must be between either transient or permanent positive energy particles.

3.5. Self-consistency requirements in a homogeneous vacuum. According to Eq. (55) the mass assigned to a vacuum particle on the basis of the fluctuation energy necessary for pair creation would increase as the vacuum density increases if the kinetic energy of the excited pair is fixed. However, according to Eq. (17) the mass defined by the gravitational attraction between two identical particles decreases as the vacuum density increases. The purpose of this and the following two sections is to show that the relations between mass and vacuum density are in fact self-consistent. The discrepancy between mass as defined by fluctuation and by gravitational force is taken up either by the kinetic energy of transient pairs, eliminated by variation in both mass and light velocity in the presence of mass, or eliminated by indirect variations in the vacuum density. The chief features of the analysis are:

i. For vacuum particles mediating the gravitational field but not the electromagnetic field the energy-defined gravitational mass is always equal to the fluctuation energy required for pair creation. The gravitational mass of vacuum particles mediating both the electromagnetic and gravitational fields must be less than the fluctuation energy, with the difference being absorbed in the positive kinetic energy of the pairs mediating the electromagnetic field.

ii. The mass of all particles increases in the presence of a depression in vacuum density (which we eventually show is equivalent to the presence of mass). But this increase is compensated by a decrease in light velocity. This is what allows the fluctuation density to control the gravitational force in a manner which is consistent with the mass as defined by the gravitational force.

iii. The fluctuation energy required for initiating skipping changes with the vacuum density. But at the same time the skipping energy must remain constant throughout the skipping process. To satisfy these dual requirements in a vacuum of variable density the skipping process must induce an accompanying depression-compression wave in the vacuum density surrounding it.

The alteration of mass and light velocity by gravitational fields is qualitatively compatible with that required by general relativity. The depression-compression waves provide the link between virtual radiation and radiation carrying real energy (to be discussed in Section 3.7).

3.5.1. Gravitational mass of vacuum particles mediating the gravitational field. Let m_0^g denote the rest mass of vacuum particles mediating the gravitational field (but not the electromagnetic field) as determined by the minimum fluctuation energy required for pair production. The total relativistic energy of the promoted particles is $\Delta E_g = 2m_0^g c^2 + T_0^g$, where ΔE_g denotes the fluctuation energy associated with the gravitational field and T_0^g denotes the kinetic energy when the mass is defined in terms of the minimum fluctuation energy required for pair production. Let m_g^g denote the rest mass of the vacuum particles mediating the gravitational field (but not the electromagnetic field) as determined by the gravitational interaction between two such particles (Eq. 17). The total relativistic energy of the promoted particles in terms of the gravitational rest mass is given by $\Delta E_g = 2m_g^g c^2 + T_g^g$, where T_g^g denotes the kinetic energy when the rest mass is defined in this manner. Gravitational vacuum particles have the following two properties:

a. In the absence of constraints $T_0^g = T_g^g = 0$.

b. In the absence of constraints $m_0^g = m_g^g$ (assuming constant light velocity).

To prove (a) equate the two expressions for ΔE_g (i.e. the energy required for pair production in terms of m_0^g and m_g^g). This gives

$$m_0^g c^2 + T_0^g = m_g^g c^2 + T_g^g \tag{71}$$

If these can vary independently and are always equal to each other, ΔE_g must be constant. If $T_0^g > 0$, pairs could be created for energies $m_0^g c^2 < \Delta E_g$, contradicting the constancy of ΔE_g. A similar argument holds if $T_g^g > 0$. But as a more instructive argument, suppose that T_g^g fluctuates to a value greater than zero. Then $\Delta E_g \sim \hbar/\tau_{\text{grav}}$ would be smaller, implying an increase in τ_{grav} and a consequent corrective increase in F_{grav} and m_g^g. This relationship is explicit in Eq. (17) since m is now m_g^g and ρ_{grav} decreases as τ_{grav} increases.

Property (b) follows directly from (a), assuming that the appearance of positive energy mass due to the excitation of a vacuum particle does not alter light velocity. If light velocity were altered from c_0 to a smaller value c_g, this would require m_g^g to be larger than m_0^g if we set the energy conservation requirement as $m_0^g c_0^2 = m_g^g c_g^2$. According to general relativity, light velocity is altered slightly by the presence of mass. We shall see in Section 3.6 that such an alteration is actually required for the internal consistency of the vacuum particle interpretation.

3.5.2. Gravitational mass of vacuum particles mediating the electromagnetic field. Let m_0^{em} denote the rest mass of vacuum particles mediating the electromagnetic field as determined by the minimum fluctuation energy required for pair production. The total relativistic energy of the promoted particles will be denoted by $\Delta E_{em} = 2m_0^{em}c^2 + T_0^{em}$, where T_0^{em} denotes the kinetic energy of the pair when it is mediating this field and when its mass is defined in terms of the minimum energy required for pair production. Let m_g^{em} denote the rest mass of vacuum particles mediating the electromagnetic field as determined by the gravitational interaction between two such particles (Eq. 17). The total relativistic energy of promoted particles in terms of this gravitational rest mass is $\Delta E_{em} = 2m_g^{em}c^2 + T_{em}^{em}$, where T_{em}^{em} denotes the kinetic energy of the pair when it is mediating the electromagnetic field and when its mass is defined in terms of its gravitational interactions.

The new feature is that vacuum particles mediating the electromagnetic field should also mediate the gravitational field. Since the fluctuation density of the electromagnetic field is smaller than the fluctuation density of the gravitational field, m_0^{em} must be smaller than m_0^g. For these particles we must write

$$\Delta E_g = 2m_g^{em} + T_g^{em} \tag{72}$$

where T_g^{em} denotes the kinetic energy of an electromagnetic carrier when it is recruited for a skipping process mediating the gravitational field. Electromagnetic carriers have the following important properties:

a. $T_g^{em} > T_{em}^{em}$ and can assume its smallest possible value when $T_{em}^{em} = T_0^{em} = 0$ (with $m_g^{em} = m_0^{em}$).

b. T_g^{em} actually assumes its smallest possible value when electromagnetic vacuum particles are recruited for mediating the gravitational field.

c. m_g^{em} assumes the largest possible value compatible with $m_g^g - m_g^{em} = T_g^{em}/2c^2$, where $T_g^{em} > 0$. Larger values of m_g^{em} are incompatible with the existence of the electromagnetic field.

d. The equilibrium between the gravitational attraction and electromagnetic repulsion among vacuum particles provides a constraint which usually prevents m_g^{em} from increasing to a value greater than allowed by condition (b).

e. The assumption that $m_g^{em} = m_0^{em}$ is also justified by the equilibrium between gravitational and electromagnetic forces (assuming constant light velocity).

f. Electric charge must always be associated with mass.

To prove (a), first note that the proof of property (a) in 3.5.1 above can be carried over to show $m_g^{\text{em}} = m_0^{\text{em}}$. Also note that since $\Delta E_g > \Delta E_{\text{em}}$, we can write $2m_g^{\text{em}}c^2 + T_g^{\text{em}} > 2m_0^{\text{em}}c^2 + T_{\text{em}}^{\text{em}}$. Thus $T_g^{\text{em}} > T_{\text{em}}^{\text{em}}$. Since $m_g^{\text{em}} = m_0^{\text{em}}, T_{\text{em}}^{\text{em}} = T_0^{\text{em}}$. Under this condition T_g^{em} can assume its smallest possible value when $T_{\text{em}}^{\text{em}} = T_0^{\text{em}} = 0$ (since negative kinetic energies are excluded).

To prove (b), suppose that T_g^{em} fluctuates to a value larger than its minimum possible value. Then it would be possible for transient pair production to occur with smaller values of $\Delta E_g \approx \hbar/\tau_{\text{grav}}$, leading to an increased gravitational force and therefore to a corrective increase in m_g^{em}. To calculate the value of m_g^{em} in terms of m_g^g, note that (because of 3.5.1a) we can write $\Delta E_g = 2m_g^g c^2$. Combining this with Eq. (72) gives $m_g^g - m_g^{\text{em}} = T_g^{\text{em}}/2c^2$, where we can again identify m_g^{em} with m_0^{em}. Since $\Delta E_{\text{em}} < \Delta E_g, m_g^{\text{em}} < m_0^g (= m_g^g)$. Thus $T_g^{\text{em}} > 0$. Suppose that m_g^{em} assumes a larger value than allowed by this equation, to be denoted by $\hat{m}_g^{\text{em}}$. Since $T_{\text{em}}^{\text{em}} = 0$, we now have $\Delta E_{\text{em}} < 2\hat{m}_g^{\text{em}}c^2$. But this means that the mass of the electromagnetic vacuum particles has grown too large for electromagnetic skipping to occur. That is, the magnitude of the energy required for pair production is now conjugate to a fluctuation length which is shorter than the average distance between vacuum particles. When this occurs the electromagnetic field is turned off.

The key point in this argument is that the pair production processes with higher than the minimum value of T_g^{em} do not contribute to propagative skipping. Only those with the minimum value contribute to the gravitational field, with deviations to higher values leading to a corrective increase in m_g^{em}. When particles have a value of m_g^{em} which is greater than the maximum value compatible with an electromagnetic field they still contribute to the gravitational field. As a consequence there are no second order effects on m_g^g, even though the electromagentic field disappears.

To establish (d), note that if in some region m_g^{em} increases to the point where the electromagentic field is turned off, the repulsive forces among the electromagentic vacuum particles in states of trapped skipping in that region will disappear. The fluctuation density, ρ_{grav}, will increase, assuming that this leads to a compression of the region. As ρ_{grav} increases, ΔE_g increases, so the gravitational force and therefore the gravitational mass of all vacuum particles decreases. As soon as the decrease in m_g^{em} is sufficient to satisfy condition (b) the electromagnetic field is turned on again, leading to an expansion of the vacuum due to the repulsions among those vacuum particles which are electromagnetically charged. The only dynamically stable situation is the one in which property (b) holds.

To establish property (e), suppose that m_0^{em} is less than m_g^{em}. This contradicts the fundamental requirement that pair production conform to the equivalence of mass and energy. Suppose that m_0^{em} is greater than m_g^{em}. Then the pairs could be produced with less energy, increasing the strength of the repulsive forces among electromagnetic vacuum particles and therefore expanding the vacuum. But expansion of the vacuum would result in a stronger gravitational force, leading to an increase in m_g^{em}. Thus $m_g^{em} = m_0^{em}$ at the equilibrium, justifying the identification of mass as defined by gravitation and as defined by fluctuation in the case of electromagnetic vacuum particles. This also explains why electromagnetic charge cannot exist without being attached to mass (property f). Mass is controlled by fluctuation density (or, equivalently, vacuum density). In order for vacuum particles (whether electromagnetic or purely gravitational) to have zero mass they would have to have infinite vacuum density (or equivalently, infinitesimal fluctuation intervals). But this contradicts the exclusion principle. Massless bosons cannot have electric charge since they are constructed from transient pair formation processes, therefore are composites of a particle and its antiparticle.

The question arises as to why low energy fluctuations of electromagnetic vacuum particles cannot mediate the gravitational field. If this were possible the attractive force of gravitation would become as strong as the repulsive force of electromagnetism, leading to an increase in the gravitational mass of all vacuum particles. But then the energies required for pair production would be greater than the fluctuation energies defined by the the vacuum densities, therefore propagative skipping would be terminated and both the electromagnetic and gravitational fields would disappear. The gravitational masses of all vacuum particles would then fall to zero and the field would be turned on again, reinstituting the bootstrapping process. Thus for the bootstrapping process to stabilize it is necessary for all the low fluctuation energy interactions mediated by the electromagnetic vacuum particles to be repulsive.

We note that the density of negative energy particles even in a very dilute vacuum is high in comparison to the number of positive energy particles. For all practical purposes positive energy particles may therefore be viewed as absorbers which are continually being bombarded by propagative skipping processes originating in the vacuum. If transient pair production occurs in the neighborhood of a positive energy particle the absorption process has the effect of initiating a skipping process. This is the reason why fluctuations with energy lower than that required for pair creation do not mediate direct interactions between positive energy particles or vacuum particles in a state of trapped skipping. If propagative skipping processes originating elsewhere in the universe are absorbed this leads to their termination. Furthermore, we note that all the propagative skipping processes originating in low energy fluctuations of electromagnetic vacuum particles (that is, fluctuations for which $T_0^{em} = 0$) must have a spin character which makes them repulsive.

3.6. Relation between mass and vacuum density. The analysis of gravitational mass and its relation to fluctuation mass is incomplete in an important respect. We have not so far considered the effect of space-time variations in the vacuum densities. The bootstrapping argument shows that the correspondence between vacuum particle mass and vacuum density is a consequence of the requirement for self-consistency. Over sufficiently small regions of space-time the requirement for self-consistency leads to definite relations among the different types of mass. The purpose of this section is to derive these relations and then to calculate how they are altered when the variations in vacuum density become significant.

3.6.1. Dependence of vacuum density on mass and charge. If F_{grav} is taken as Newton's inverse square law of gravitation, we can write

$$M \approx \frac{\hat{A}_{grav}}{mGc^2\hbar^2 \rho_{grav}^{\frac{4}{3}}} \qquad (73)$$

where M and m are the masses of two interacting particles. An analogous equation holds for charge if F_{Coul} is taken as Coloumb's inverse square law. If $M = m$ we have Eq. (17). The occurrence of both a positive and negative sign in this equation appears to correspond to the occurrence of positive and negative energy particles in the vacuum particle interpretation. But in fact vacuum particles only serve as absorbers when they transiently enter a positive energy state (due to the trapped skipping mechanism). From this mechanistic point of view gravitational mass is strictly speaking a positive energy quantity. Quantities such as m_g^g and m_g^{em} are most suitably interpreted as inherently positive, equal to the fluctuation defined quantities m_0^g and m_0^{em} only to an approximation which depends on the effect of mass on light velocity. For simplicity we will retain only the positive sign in the following argument, though the formal existence of the negative sign should not be forgotten.

So far we have concentrated on the fact that ρ_{grav} controls F_{grav}. But a change in M or m must also alter ρ_{grav}. This is due to the fact that positive energy objects can have very different masses. Clearly these variations cannot be absorbed by constants on the right hand side of Eq. (73). These "constants" could at most undergo small changes in magnitude. Since the total charge on an object can also vary, analogous reasoning leads to the conclusion that charge alters ρ_{em}.

Such variation in the vacuum densities cannot be uniformly spread over space-time. This would contradict the manifest fact that M is not necessarily equal to m, but that exchanging these two quantities in Eq. (73) leads to a different value of ρ_{grav}. The space-time structure of the vacuum density must thus depend on the distribution of positive energy mass and charge. To express this dependence formally in the case of two interacting particles of mass m we will write ρ_{grav} as $\rho_g[m, r_1, r_2, s]$, where s specifies the location in space at which the density is measured, r_1 and r_2 specify the positions, and square brackets will always be used to indicate functional notation. In fact ρ_{grav} must depend on all the mass in the universe, but for simplicity we will regard this as a constant background. The electromagnetic vacuum density ρ_{em} must be written as $\rho_{em} = \rho_{em}[m, q, r_1, r_2, s]$, where we again consider only two interacting particles, in this case of equal charge as well as of equal mass. That the gravitational vacuum density can be written as $\rho_g[m, r_1, r_2, s]$ rather than as $\rho_g[m, q, r_1, r_2, s]$ is due to the fact that ρ_{grav} is always many orders of magnitude larger than ρ_{em} (see Section 2.4). Thus the fraction of the total number of vacuum particles that can be affected by a change in q is too negligible to lead to any measurable effects on the structure of ρ_{grav}. As a consequence the electromagnetic field has no measurable effect on m_g^g, the gravitational mass of vacuum particles mediating only the gravitational field. The reverse is not the case. Since the gravitational field affects all vacuum particles it must affect ρ_{em} as well as ρ_{grav}. The dependence of ρ_{em} on charge is much more local than the dependence of either ρ_{grav} or ρ_{em} on mass, due to the fact that the overall charge distribution of positive energy particles in the universe must be neutral.

Space-time variations in the vacuum density must actually lead to field structures slightly different from the classical ones (since the inverse square law is masked by variations in the vacuum density). This masking corresponds to a screening effect in the case of the electromagnetic field and to slight deviations from the Newtonian law in the case of the gravitational field. But for the present purposes the classical Newtonian law of gravity is sufficient for analyzing the effect of variations in vacuum density on the skipping process. These effects are independent of the precise form of the gravitational law (though this form would of course be necessary for calculating the vacuum density structure itself).

3.6.2. Equilibrium equations for vacuum particle mass and density. Consider two particles of mass $m = M_g$. The introduction of these two particles alters the surrounding vacuum density. As a first approximation we assume that the alteration in vacuum density is uniform throughout the universe. This is equivalent to the assumption that the Newtonian law of gravitation is sufficient for the analysis. The formal expression of this assumption is that $\partial \rho_g / \partial r_{1_i} = 0, \partial \rho_g / \partial r_{2_i} = 0$, and $\partial \rho_g / \partial s_i = 0$ for all i (where r_{j_i} is the ith component of $\mathbf{r}_j$). When this assumption obtains we will write $\rho_g[M_g, \mathbf{r}_1, \mathbf{r}_2, \mathbf{s}]$ as $\rho_g[M_g]$. This notation is intended to represent the circumstance that M_g can have any value and is therefore appropriately regarded as the independent variable. In this approximation ρ_{grav} is controlled by the two particles of mass M_g and by an implicit background mass whose distribution is viewed as effectively constant.

We can now summarize the formulas which control mass when the gravitational masses of vacuum particles have reached equilibrium. For two interacting positive energy particles of mass M_g Eq. (73) may be abbreviated as

$$M_g \approx \tilde{A}_{\text{grav}} \rho_g[M_g]^{-\frac{2}{3}} \tag{74a}$$

where $\tilde{A}_{\text{grav}} = A_{\text{grav}}^{\frac{1}{2}} |<f|V_{\text{grav}}|i>| G^{\frac{1}{2}} c\hbar$ and we consider only the positive root. For two vacuum particles in a state of trapped skipping and of mass m_g^g, Eq. (73) should be written as

$$m_g^g[\rho_g[M_g, m_g^g]] \approx \tilde{A}_{\text{grav}} \rho_g[M_g . m_g^g]^{-\frac{2}{3}} \tag{74b}$$

Here we assume that the two permanent positive energy particles of mass M_g are still present. But the additional dependence of ρ_{grav} on m_g^g is necessary since the density of vacuum particles available for skipping is altered by the entry of the two vacuum particles into a state of trapped skipping. If the two permanent positive energy particles are ignored $\rho_g[M_g, m_g^g]$ may be replaced by $\rho[m_g^g]$. The functional notation $m_g^g[\rho_g(M_g, m_g^g)]$ will be used to emphasize the fact that vacuum particle mass of purely gravitational vacuum particles is controlled by ρ_{grav}, and that ρ_{grav} is altered by positive energy mass. Eq. (74b) assumes that m_g^g is a definite property of a particle in a state of trapped skipping and that the extent to which the virtual skipping energy exceeds the virtual energy required for propagative skipping need not be considered due to the fact that the particle is confined to a fluctuation volume. The occurrence of a particle in a state of trapped skipping entails the occurrence of a trapped hole, but as the hole and the particle are confined to the same fluctuation volume, the effect on vacuum density can be thought of in terms of the particle alone. Eq. (74b) also holds for the pairwise interaction between two permanently promoted positive energy particles of mass m_g^g, again assuming the background presence of the mass M_g. The alteration in vacuum density associated with two particles in a state of trapped skipping is equal to the alteration produced by the permanent promotion of one vacuum particle (since in this case a distinct antiparticle is also produced).

In the case of electromagnetic vacuum particles, Eq. (73) becomes

$$m_g^{\text{em}}[\rho_g[M_g, m_g^{\text{cm}}]] \approx \tilde{A}_{\text{grav}}\rho_g[M_g, m_g^{\text{em}}]^{-\frac{2}{3}} \qquad (74c)$$

Since $m_g^{\text{em}} < m_g^g$, this means that trapped skipping of purely gravitational vacuum particles is associated with a larger depression of vacuum density than trapped skipping of electromagnetic vacuum particles, consonant with the fact that particles of greater mass should in some sense be larger than particles of lesser mass. The only alternative to this would be for the constant, $\tilde{A}_{\text{grav}}$, to differ in Eqs. (73) and (74), say due to a difference in V_{grav} in the two cases.

As above, the notation $m_g^g[\rho_g[M_g, m_g^g]]$ and $m_g^{\text{em}}[\rho_g[M_g, m_g^{\text{em}}]]$ expresses the fact that m_g^g and m_g^{em} are here being viewed as dependent variables which must be subject to the equation for rest mass in terms of the energy required for pair production. For purely gravitational vacuum particles this equation may be written (from Eq. 55)

$$m_0^g[\rho_g[M_g]] \approx \frac{\hbar\rho_g[M_g]^{\frac{1}{3}}}{2c} \qquad (75a)$$

The kinetic energy term $(T_- - T_+)/2c^2$ has been set equal to zero since $T_+ = T_0^g$, which is equal to zero whenever the pair production occurs as part of a propagative skipping process (cf. 3.5.1). T_- can always be set to zero, as discussed in Section 3.3. The controlling vacuum density could be written as ρ_{grav} since the vacuum particle is initially in its negative energy state. When it is excited to a positive energy state, either transiently or permanently, the vacuum density must be altered by the positive energy mass. We defer (to the next section) considering the effect of this alteration since it cannot alter the energy required for pair production.

For electromagnetic vacuum particles Eq. (55) becomes

$$m_0^{\text{em}}[\rho_{\text{em}}[M_g, Q]] \approx \frac{\hbar \rho_{\text{em}}[M_g, Q]^{\frac{1}{3}}}{2c} \tag{75b}$$

where Q is the charge associated with M_g (if any). As with $\rho_{\text{grav}}, \rho_{\text{em}}$ is here treated as a constant background affected only by M_g and Q. The kinetic energy term, T_0^{em}, is again zero since m_0^{em} is expressed in terms of its corresponding vacuum density ρ_{em} (see Section 3.5.2, properties a and b). To express it in terms of ρ_{grav} note that according to property b of Section 3.5.2

$$m_0^g[\rho_g[M_g]] - m_0^{\text{em}}[\rho_{\text{em}}[M_g, Q]] = \frac{T_g^{\text{em}}}{2c^2} \tag{76}$$

where $m_0^g[\rho_g[M_g]] = m_g^g$ and $m_0^{\text{em}}[\rho_{\text{em}}[M_g, Q]] = m_g^{\text{em}}$. Combining Eq. (76) with (75b)

$$m_0^g[\rho_g[M_g]] \approx \frac{\hbar \rho_{\text{em}}[M_g, Q]^{\frac{1}{3}}}{2c} + \frac{T_g^{\text{em}}}{2c^2} \tag{77a}$$

or with Eq. (75a)

$$m_0^{\text{em}}[\rho_{\text{em}}[M_g, Q]] \approx \frac{\hbar \rho_g[M_g]^{\frac{1}{3}}}{2c} - \frac{T_g^{\text{em}}}{2c^2} \tag{77b}$$

This confirms the fact that m_0^{em} and ρ_{grav} can be altered independently and that this independence is based on the possibility of altering the T_g^{em} term. Eq. (77b) is a direct restatement of Eq. (55) under a condition in which the kinetic energy term cannot go to zero.

The reason for carrying the rather cumbersome functional notation can now be made clear. The masses m_g^g and m_0^g are not the same in Eqs. (74b) and (75a); nor are the masses m_g^{em} and m_0^{em} the same in Eqs. (74c) and (77b). If these masses were the same the first pair of equations could be combined to obtain

$$\frac{\hbar \rho_g[M_g]^{\frac{1}{3}}}{2c\rho_g[M_g, m_g^g]^{-\frac{2}{3}}} \approx \tilde{A}_{\text{grav}} \tag{78a}$$

while the second pair could be combined to obtain

$$\frac{\hbar \rho_g[M_g]^{\frac{1}{3}}}{2c\rho_g[M_g, m_g^{\text{em}}]^{-\frac{2}{3}}} - \frac{T_g^{\text{em}}}{2c^2\rho_g[M_g, m_g^{\text{em}}]^{-\frac{2}{3}}} \approx \tilde{A}_{\text{grav}} \qquad (78b)$$

But $T_g^{\text{em}} > 0$ (according to properties a and b of Section 3.5.2). Thus both of these equations can hold only if $\rho_g[M_g, m_g^g] > \rho_g[M_g, m_g^{\text{em}}]$. This is why the rather subtle feature that the vacuum density controlling the force between two vacuum particles in a state of trapped skipping is different than the vacuum density when these particles are in a purely negative energy state is crucial for the logical structure of the theory.

   3.6.3. Effect of mass on light velocity. We can now use a simple energy argument to show that the vacuum particle interpretation yields the result that light velocity decreases in the presence of positive energy mass. Consider a vacuum particle of mass m_0^g which is promoted to a positive energy particle and hole, each of mass m_g^g. Since the issue to be investigated is whether light velocity depends on mass (or, alternatively, on ρ_{grav}, we shall write c as $c[\rho_{\text{grav}}]$ and understand that c in all the previously derived equations should be replaced by this functional form. Since we no longer assume that c is constant, we can no longer assume that $m_0^g = m_g^g$ or that $m_0^{\text{em}} = m_g^{\text{em}}$ (properties 3.5.1-b and 3.5.2-c).

   Recall (from Eq. 52) that according to the usual interpretation of the Dirac vacuum $\Delta E = E_+ - E_-$, where E_+ is the energy of a promoted particle in its lowest positive energy state (that is, with the kinetic energy T_+ equal to zero), and E_- is the energy of the particle in its highest negative energy state. Since $E_+ = E_-$, we can write $\Delta E = 2E_+$. Both the positive energy particle and the hole produced by the promotion have energy E_+. The new feature is that the vacuum density is different before and after promotion. Thus the energy required for promotion (say the fluctuation energy required for transient promotion) depends on the initial vacuum density and is therefore given by

$$\Delta E_g = 2m_0^g[\rho_g[M_g]]c[\rho_g[M_g]]^2 \qquad (79)$$

where $\Delta E = \Delta E_g$ in the case of gravitational vacuum particles. The rest energy subsequent to promotion is defined by force as determined by the new vacuum density, therefore is given by

$$2E_{g+} = 2m_g^g[\rho_g[M_g, m_g^g]]c[\rho_g[M_g, m_g^g]]^2 \qquad (80)$$

where E_+ will be denoted by E_{g+} in the case considered. Substituting Eq. (75a) in Eq. (79) and Eq. (74b) into Eq. (80) and equating the resulting expressions for ΔE and $2E_+$

$$\frac{Dc[\rho_g[M_g, m_g^g]]}{c[\rho_g[M_g]]} \approx \rho_g[M_g]^{\frac{1}{3}} \rho_g[M_g, m_g^g]^{\frac{2}{3}} \tag{81}$$

where $D = 2A_{\text{grav}}^{\frac{1}{2}} | < f|V_{\text{grav}}|i > |G^{-\frac{1}{2}}\hbar^{-2}$. Here again it is clear that the negative root would not be meaningful since the vacuum densities are both positive and the ratio of light velocities is positive. Since D is constant, $c[\rho_g[M_g, m_g^g]]$ must decrease if $\rho_g[M_g, m_g]$ decreases. But $\rho_g[M_g, m_g^g]$ decreases when m_g^g increases, thereby establishing that c decreases in the vicinity of positive energy mass.

The equality of Eqs. (79) and (80) can be used to obtain an equilibrium equation for m_g^g. Combining Eq. (74b) and (75a) to eliminate $\hbar$,

$$m_g^g[\rho_g[M_g, m_g^g]]m_0^g[\rho_g[M_g]] \approx \frac{\hat{A}_{\text{grav}}^{\frac{1}{2}} \rho_g[M_g]^{\frac{1}{3}}}{2G^{\frac{1}{2}}c[\rho_g[M_g, m_g^g]]c[\rho_g[M_g]]\rho_g[M_g, m_g^g]^{\frac{2}{3}}} \tag{82}$$

Using the equality of ΔE_g and $2E_{g+}$ to eliminate $m_0^g[\rho_g[M_g]]$

$$m_g^g[\rho_g[M_g, m_g^g]] \approx \frac{\hat{A}_{\text{grav}}^{\frac{1}{4}} \rho_g[M_g]^{\frac{1}{6}} c[\rho_g[M_g]]^{\frac{1}{2}}}{2^{\frac{1}{2}}G^{\frac{1}{4}}\rho_g[M_g, m_g^g]^{\frac{1}{3}}c[\rho_g[M_g, m_g^g]]^{\frac{3}{2}}} \tag{83}$$

However, to investigate how ΔE_g is controlled by vacuum density and light velocity it is more convenient to use the formula

$$m_0^g[\rho_g[M_g]] \approx \frac{\hat{A}_{\text{grav}}^{\frac{1}{4}} \rho_g[M_g]^{\frac{1}{6}} c[\rho_g[M_g, m_g^g]]^{\frac{1}{2}}}{2^{\frac{1}{2}}G^{\frac{1}{4}}\rho_g[M_g, m_g^g]^{\frac{1}{3}}c[\rho_g[M_g]]^{\frac{3}{2}}} \tag{84}$$

which is obtained by eliminating $m_g^g[\rho_g[M_g, m_g^g]]$ from Eq. (82). Combining with Eq. (81)

$$m_0^g[\rho_g[M_g]]c[\rho_g[M_g]] \approx \frac{G^{\frac{1}{4}}\hbar^2 \rho_g[M_g]^{\frac{1}{3}}}{2^{\frac{3}{2}}\hat{A}_{\text{grav}}^{\frac{1}{4}}} \tag{85}$$

Since the right hand side decreases as $\rho_g[M_g]$ decreases, and since $c[\rho_g[M_g]]$ decreases as $\rho_g[M_g]$ decreases, the fluctuation energy $\Delta E_g = 2m_0^g[\rho_g[M_g]]c[\rho_g[M_g]]^2$ must also decrease. Since $\Delta E_g = 2E_{g+}$, this means that the rest energy of vacuum particles is dominated by the fluctuation energy required for pair production, but at the same time the increase in m_g^g with decreasing vacuum density implies that the rest mass per se is dominated by gravitational force. Whether or not the formal rest mass quantity m_0^g can ever decrease with decreasing vacuum density depends on whether the decrease in $\rho_g[M_g]$ in Eq. (85) can dominate the decrease in $c[\rho_g[M_g]]$ at low vacuum densities.

Now it is possible to reexamine how it is possible for the mass as defined by the energy required for pair production (m_0^g) and the mass as defined through gravitational force (m_g^g) to have an opposite dependence on vacuum density (Eqs. 75a and 74b). As vacuum density decreases, m_g^g increases, as can be seen from the fact that the denominator of Eq. (83) depends more strongly on vacuum density than does the numerator. However, according to Eq. (81) this increase in m_g^g must be accompanied by a decrease in $c[\rho_g[M_g, m_g^g]]$ and of course a larger decrease in $c[\rho_g[M_g, m_g^g]]^2$. This decrease in light velocity in the vicinity of a promoted particle and antiparticle allows the energy required for pair production as defined by the gravitational mass to match the energy required for pair production as defined by the vacuum densities. At very low vacuum densities the shift in light velocity with pair production becomes more significant (since m_g^g becomes larger), leading to a greater discrepancy between m_g^g and m_0^g.

The exact relation between mass and light velocity is still undetermined. Calculating this dependence from the assumptions of the theory is not possible at this point since we have yet to determine the relation between vacuum density and mass. The assumption in this regard has been that the Newtonian approximation that the depression in vacuum density induced by mass is uniform throughout space-time is sufficient at this stage. It is possible to remove this simplification; but in the meantime it is sufficient to use the relativistic dependence of mass on light velocity to obtain some information about the dependence of vacuum density on mass. If $c[\rho_g[M_g]]$ is taken as the light velocity in free space and $\rho_g[M_g]$ as the gravitational vacuum density in free space Eq. (81) reduces to

$$\rho_g[M_g, m_g^g] \propto c[\rho_g[M_g, m_g^g]]^{\frac{3}{2}} \qquad (86)$$

According to general relativity coordinate light velocity decreases with gravitational potential, with a small retardation in the transverse direction and a smaller retardation in the radial direction [18]. According to Eq. (86) gravitational vacuum density is affected slightly more strongly than is light velocity. But the effect must be different in the radial and transverse directions and must decrease with distance from mass (since the gravitational potential is inversely proportional to r).

3.5.4. Extension of the analysis to m_g^{em}. In the case of electro-magentic vacuum particles the energy required for pair production is given by

$$\Delta E_{em} = 2m_0^{em}[\rho_{em}[M_g, Q]]c[\rho_g[M_g]]^2 \tag{87}$$

The rest energy subsequent to promotion is given by

$$2E_{em+} = 2m_g^{em}[\rho_g[M_g, m_g^{em}]]c[\rho_g[M_g, m_g^{em}]]^2 \tag{88}$$

Combining Eqs. (74c) and (75b) to eliminate $\hbar$ and using the equality of Eqs. (87) and (88) to eliminate $m_0^{em}[\rho_{em}[M_g, Q]]$ yields

$$m_g^{em}[\rho_g[M_g, m_g^{em}]] \approx \frac{\hat{A}_{grav}^{\frac{1}{4}}\rho_{em}[M_g, Q]^{\frac{1}{6}}c[\rho_g[M_g]]^{\frac{1}{2}}}{2^{\frac{1}{2}}G^{\frac{1}{4}}\rho_g[M_g, m_g^{em}]^{\frac{1}{3}}c[\rho_g[M_g, m_g^{em}]]^{\frac{3}{2}}} \tag{89}$$

where the equations have been corrected to incorporate the appropriate dependence of light velocity on vacuum densities. This is analogous to Eq. (83), the equilibrium equation for m_g^g. Combining these two equilibrium equations

$$\frac{m_g^{em}[\rho_g[M_g, m_g^{em}]]}{m_g^g[\rho_g[M_g, m_g^g]]} \approx \frac{\rho_{em}[M_g, Q]^{\frac{1}{6}}\rho_g[M_g, m_g^g]^{\frac{1}{3}}c[\rho_g[M_g, m_g^g]]^{\frac{3}{2}}}{\rho_g[M_g]^{\frac{1}{6}}\rho_g[M_g, m_g^{em}]^{\frac{1}{3}}c[\rho_g[M_g, m_g^{em}]]^{\frac{3}{2}}} \tag{90}$$

The point to note is that the terms in the numerator are all smaller than the corresponding terms in the denominator, and as a consequence $m_g^g[\rho_g[M_g, m_g^g]]$ must in general be many orders of magnitude larger than $m_g^{em}[\rho_g[M_g, m_g^{em}]]$. To study the kinetic energy term equate (87) and (88) and combine with Eq. (77b) to obtain

$$m_g^{em}[\rho_g[M_g, m_g^{em}]]c[\rho_g[M_g, m_g^{em}]]^2 \approx \frac{\hbar\rho_g[M_g]^{\frac{1}{3}}c[\rho_g[M_g]]}{2} - T_g^{em} \tag{91}$$

where Eq. (77b) is again corrected to incorporate the dependence of c on vacuum density. The first term on the right hand side of Eq. (91) decreases when $\rho_g[M_g]$ decreases (since $c[\rho_g[M_g]]$ changes in the same direction as $\rho_g[M_g]$). Thus the fluctuation energy for electromagnetic skipping must decrease as ρ_{grav} decreases, just as for gravitational vacuum particles. Similarly m_g^{em} must increase since $c[\rho[M_g, m_g^{em}]]$ decreases. The difference from the purely gravitational case is the T_g^{em}

term, which must increase as the difference between ρ_{grav} and ρ_{em} becomes greater. Thus for any given value of ρ_{grav}, $m_g^{em}[\rho_g[M_g, m_g^{em}]]$ will be higher when ρ_{em} is higher. This again illustrates that it is the kinetic energy term which allows ρ_{em} to vary independently of ρ_{grav}.

The analog of Eq. (81) may be obtained by inserting Eq. (75b) into Eq. (87) and Eq. (74c) into Eq. (88). Equating the resulting expressions

$$\frac{Dc[\rho_g[M_g, m_g^{em}]]}{c[\rho_g[M_g]]} \approx \rho_{em}[m_g, Q]^{\frac{1}{3}} \rho_g[M_g, m_g^{em}]^{\frac{2}{3}} \qquad (92)$$

If $\rho_{em}[M_g, Q]$ decreases, m_g^{em} decreases (since T_g^{em} increases). This compensating increase is important since it buffers the effect of light velocity from changes in the electromagnetic vacuum density, consonant with the fact that light velocity is not noticeably affected by electromagnetic phenomena.

Combining Eq. (92) with Eq. (81)

$$\frac{c[\rho_g[M_g, m_g^{em}]]}{c[\rho_g[m_g, m_g^g]]} \approx \frac{\rho_{em}[M_g, Q]^{\frac{1}{3}} \rho_g[M_g, m_g^{em}]^{\frac{2}{3}}}{\rho_g[M_g]^{\frac{1}{3}} \rho_g[M_g, m_g^g]^{\frac{2}{3}}} \qquad (93)$$

Combining with Eq. (90)

$$\frac{m_g^{em}[\rho_g[M_g, m_g^{em}]]}{m_g^g[\rho_g[M_g, m_g^g]]} \approx \frac{c[\rho_g[M_g, m_g^g]]}{c[\rho_g[M_g, m_g^{em}]]} \qquad (94)$$

Thus the ratio of electromagnetic and gravitational vacuum particle masses is the reciprocal of the ratio of light velocities in the neighborhood of these masses. The relatively small mass associated with permanently promoted electrons means that particle promotion has only a small effect on light velocity in this case. By contrast the very much larger masses associated with gravitational vacuum particles would produce dramatic reductions in light velocity and vacuum density. But these masses are so large that sufficient energy might never be available for permanently promoting such particles under conditions ordinarily accessible to observation. Under all ordinary conditions they could only exist either in the vacuum state or in transiently excited states of extremely short lifetime.

3.7. *Wave aspect of vacuum particle skipping.* The variation in vacuum density in the neighborhood of two interacting particles is too small to cause quantized force laws to significantly depart from Newton's law. However, this variation cannot be ignored from the standpoint of the skipping processes which mediate the gravitational field. The problem is that the fluctuation energy required for pair production must increase as the gravitational vacuum density increases, therefore must be greater in free space (where the gravitational field is low) than in the immediate vicinity of an absorbing particle. However, once initial pair production occurs this fixes the skipping energy, ΔE_g, for each step of the skipping process. Thus as the skipping process propagates away from the initiating particle its skipping energy becomes insufficient for pair production due to the fact that it must enter into space-time regions of higher vacuum density. Assuming that the skipping process could survive this increase in vacuum density the opposite problem of entering a region of lower vacuum density will also occur if the skipping process enters the space-time territory surrounding a more massive object. At this point the skipping energy will be too high for propagative skipping to continue. The argument applies to propagative skipping processes originating in the vicinity of trapped absorbers in "free" space as well as to skipping processes originating in the neighborhood of permanent positive energy masses.

Electromagnetic skipping processes face the same problem. The electromagnetic vacuum density is smallest in the immediate neighborhood of charge and increases as the electromagnetic field strength increases. Thus initial pair production for an electromagnetic skipping process sets a lower initial fluctuation energy for the skipping process than is required for pair production away from the initiating particle; or the fluctuation energy becomes too high for propagative skipping in a space-time region of higher electromagnetic field strength.

The variation of vacuum density with gravitational or electromagnetic field strength thus reveals a fundamental condition which must be met by skipping processes, but which is not manifest under the simplifying assumption of a completely homogeneous vacuum. There are only two possibilities for satisfying this condition:

i. The skipping energies are incremented or decremented in the course of propagative skipping to correspond to the vacuum densities.

ii. The skipping process alters the vacuum densities in the neighborhood surrounding it in such a manner as to mask vacuum inhomogeneities (therefore in such a manner as to eliminate the effects of changes in the gravitational or electromagnetic field strength).

Alternative (i) is implausible since incrementing or decrementing the skipping energy requires the presence of an absorber; but the interaction of the skipping process with an absorber results in its termination in any case. Alternative (ii) implies that skipping processes are accompanied by a wave of compression and expansion in the vacuum density, with the phase relation between skipping process and compression-expansion wave being such that the skipping energy always matches the vacuum density.

Alternative (ii) has the attractive feature that it associates a wave aspect with vacuum particle skipping. The purpose of this section is to demonstrate the existence of this wave phenomenon for both the gravitational and electromagnetic vacuum densities. By itself this existence argument does not establish the structure of these waves. However, as vacuum density waves are coordinated with virtual skipping processes mediating the $1/r^2$ fields it is reasonable to associate them as well with the $1/r$ electromagnetic and gravitational radiation fields implied by Maxwell's equations and Einstein's gravitational field equations.

3.7.1. Gravitational density waves. Density waves occur whenever the gravitational or electromagnetic field strengths are inhomogeneous, since in this case the vacuum density must be inhomogeneous. For simplicity we consider a single object of mass M and denote the spacetime dependence of the gravitational vacuum density by $\rho_g[M, \mathbf{r}]$. In reality the detailed structure of the vacuum density is determined by all other masses as well. But if M is sufficiently large and sufficiently far from all other masses these other masses can be treated as a constant background.

Referring back to Eq. (75a) and to Eq. (79), the energy required for pair production may be written as

$$\Delta E_g = 2m_0^g[\rho_g[M, \mathbf{r}]]c[\rho_g[M, \mathbf{r}]]^2 \approx \hbar\rho_g[M, \mathbf{r}]^{\frac{1}{3}}c[\rho_g[M, \mathbf{r}]] \qquad (95)$$

But in the volume surrounding a single positive energy object

$$\partial\rho_g[M, \mathbf{r}]/\partial\mathbf{r} > 0 \qquad (96a)$$

and

$$\partial c[\rho_g[M, \mathbf{r}]]/\partial\rho_g[M, \mathbf{r}] > 0 \qquad (96b)$$

These increases of vacuum density and light velocity with distance from mass can be viewed as a direct expression of the fact that the equivalence of mass and energy requires that all forms of energy have a depressing effect on vacuum density. Since the energy density of the field surrounding a massive object decreases with distance from the object the vacuum density must increase with this distance.

Taken together Eqs. (95) and (96a,b) provide the formal justification for the statement that ΔE_g cannot remain constant for a gravitational skipping process unless skipping is linked to local alterations in the gravitational vacuum density. This is evident from the fact that both $\rho_g[M, \mathbf{r}]$ and $c[\rho_g[M, \mathbf{r}]]$ change in the same direction as the skipping process either propagates away from or towards the absorbing particle. In reality of course the influence of other masses must eventually cause Eqs. (96a) and (96b) to fail when $\mathbf{r}$ becomes sufficiently large. But the argument remains valid since $\rho_g[M, \mathbf{r}]$ and $c[\rho_g[M, \mathbf{r}]]$ always change in the same direction in any case.

For the skipping processes implied by quantized force laws to occur it is therefore necessary for them to become enveloped by a propagating wave of elevated and depressed vacuum density whenever they enter a space-time territory in which the vacuum density differs from that which obtained at the point of initiation. This wave, to be called a compression-expansion wave, must have the following properties:

i. The skipping energy must remain constant from the point of initiation to point of absorption, independent of the distribution of mass in the universe. Skipping processes initiated in the neighborhood of large mass will therefore have a lower skipping energy than those initiated in the neighborhood of a smaller mass (since the vacuum density is lower in the case of a larger mass).

ii. When the skipping process enters a region of space-time in which the vacuum density is higher than at the point of initiation the skip must occur in the expanded (or depressed) region of the compression-expansion wave. When the skipping process enters a region of space-time in which the vacuum density is lower than at the point of initiation the skip must occur in the compressed region of the compression-expansion wave. In general the phase relation between the skipping process and the compression-expansion wave must be such that the advance of the skips coincide with the advance of that portion of the compression-expansion wave whose vacuum density matches the vacuum density at the point of initiation.

iii. The principle of energy conservation requires that the compression part of the density wave be equal to the expansion part and be coupled to it. Equality of these two parts eliminates the necessity of assuming the creation of additional virtual energy. Attachment of the two parts has the important consequence that no net mass (or energy) is assigned to the wave. Under conditions of equality and coupling neither the wave nor the virtual skipping process associated with it will gravitate. That local compressions of the vacuum density must entail expansions elsewhere may be viewed as equivalent to the assumption that the volume of the universe is constant. But that the compressions and expansions must be coupled is a distinctive requirement, equivalent to the statement that energy should be neither created or destroyed by a skipping process as it propagates from point of initiation to point of termination.

iv. The zero energy property is consistent with assigning an infinite wavelength to a virtual density wave. Such an infinite wavelength is compatible with the fact that the skips always propagate away from their source, yet does not contradict the classical quantum field theory restriction of a massless momentum carrier to a range of $\lambda/2$ (see 2.9.4).

3.7.2. Electromagnetic density waves. To formally establish the existence of electromagnetic density waves consider a single object of mass M and charge Q and denote the space-time dependence of the electromagnetic vacuum density by $\rho_{em}[M,Q,\mathbf{r}]$. As before, the detailed structure of the vacuum density is actually determined by the total distribution of mass and charge in the universe. But if the charged body is sufficiently far from other charged bodies, it will be the prime determinant of the vacuum density in some space-time volume surrounding it.

Referring back to Eqs. (75b) and (87), the energy required for pair production is now given by

$$\Delta E_{em} = 2m_0^{em}[\rho_{em}[M,Q,\mathbf{r}]]c[\rho_g[M,\mathbf{r}]]^2$$
$$\approx \hbar\rho_{em}[M,Q,\mathbf{r}]^{\frac{1}{3}}c[\rho_g[M,\mathbf{r}]] \tag{97}$$

where the dependence of c on vacuum density is represented explicitly. But in the volume surrounding a charged positive energy object

$$\partial\rho_{em}[M,Q,\mathbf{r}]/\partial\mathbf{r} > 0 \tag{98a}$$
$$\partial c[\rho_g[M,\mathbf{r}]]/\partial\mathbf{r} > 0 \tag{98b}$$

Condition (98a) is a consequence of the effect of charge on electromagnetic vacuum density and condition (98b) is a consequence of the effect of mass on the gravitational vacuum density. Given these two conditions, Eq. (97) implies that ΔE_{em} must change as a function of $\mathbf{r}$ unless skipping is linked to local alterations in the electromagnetic vacuum density. The argument remains valid for arbitrary distributions of charge and mass except for the very unlikely circumstance that increases in $\rho_{em}[M, Q, \mathbf{r}]$ are precisely compensated by decreases in $c[\rho_g[M, \mathbf{r}]]$.

The four properties of gravitational density waves listed in Section 3.7.1 also apply to electromagnetic density waves. Due to the fact that the gravitational vacuum density is very much larger than the electromagnetic vacuum density, gravitational density waves are negligibly affected by electromagnetic density waves (as long as they do not possess net mass). Since the electromagnetic vacuum density is altered in proportion to any alteration in the gravitational vacuum density the gravitational density waves will have a slight effect on the electromagnetic density waves. In addition this effect will be slightly modulated by the fact that c depends on ρ_{grav} in Eq. (97) just as it does in Eq. (95).

3.7.3. Density waves carrying real energy. The existence of density waves provides the link between virtual bosons and bosons which carry real energy. The difference is that the vacuum compression enveloping the skipping process is greater than the vacuum expansion to which it is coupled. It is by means of this net depression that the skipping process carries macroscopically observable momentum $p = h/\lambda$, and energy $\epsilon = h\nu$, where λ and ν are the wavelength and frequency of the density wave (cf. 2.11). At some point on the wave, in the transition from maximum density elevation to maximum density depression, a value of the vacuum density must occur which matches the original skipping energy. The phase relation between the skipping process and the density wave is determined by the relative position on the wave at which this matching density occurs. The wavelength of the density wave is determined by its energy, but the relation between fluctuation length and wavelength is subject to the constraint that the skipping excitation travels along with the matching vacuum density.

3.7.4. Effect of gravity on density waves and red shift. Gravitational fields exert two types of effects on density waves. First the propagation velocity is decreased in a gravitational field. Since the effect of mass on light velocity is a consequence of vacuum particle skipping this is independent of whether the density wave is purely virtual or whether it possesses real energy. The second effect is through the gravitational interaction between the density wave and some other body. The consequences of this effect are the same for both virtual and real density waves, but the mechanistic interpretation is different. Density waves carrying real energy correspond to net vacuum depressions, therefore are absorbers of gravitational skipping processes. This cannot be the case for purely virtual density waves, since these carry no net energy. However, the field surrounding a positive energy particle corresponds to a net vacuum depression, therefore possesses net energy and must gravitate. Since purely virtual density waves arise only in connection with such fields they will appear to be affected by a direct gravitational interaction, just as density waves carrying real energy are.

Consider a purely virtual electromagnetic skipping process (i.e., a purely virtual light ray) which is initiated in the neighborhood of a massive body, such as the sun. In the region of initiation r_{em} is larger than in free space, or than at the surface of the earth. In the interaction of two positive energy bodies the more massive body will therefore initiate skipping processes of lower energy and momentum. This is consistent with the fact that these lower energy skipping processes are more efficacious for actually transmitting the momentum responsible for the force (Section 2.3). In effect, the fluctuation length of a skipping process initiated in the neighborhood of a massive body is elongated.

If the light ray carries real energy this elongation leads to a red shift. This is due to the fact that the density wave must have a longer wavelength in order to maintain the proper phase relation with the skipping process. The real energy $\epsilon = h\nu = ch/\lambda$, carried by such a real density wave will be reduced, corresponding to the fact that the virtual energy carried by the underlying skipping process is reduced. The occurrence of the red shift of course implies the slowing of a clock in a gravitational field. This is due to the fact that the vibration frequency of an atom used to measure time must decrease if the wavelength of the emitted radiation increases.

3.7.5. Linkage between skipping processes carrying real and virtual energy. The occurrence of real and virtual density waves are always linked. An observer L, initially at rest with respect to body M will feel gravitational and possibly an electromagnetic attraction to M. Both M and the observer will undergo a change in momentum due to the occurrence of spontaneous, purely virtual skipping processes. Both bodies must therefore undergo an acceleration (even if they are pinned by forces of constraint). These accelerations in turn entail the occurrence of real radiation with a $1/r$ dependence, that is, the occurrence of skipping processes which become decorrelated from these bodies and which are accompanied by density waves which carry real energy. An observer L', initially accelerating relative to L, will initially report the occurrence of real radiation. But such an observer will have to admit that the initial acceleration of M in his coordinate system must have been due to a prior virtual exchange or to radiation traceable to a prior virtual exchange. In general the two fields, the undetached $1/r^2$ field and the detached $1/r$ field, must occur together for every observer. But since L and L' are in relative motion their initial reports (if made at the same time according to precalibrated clocks) will refer to different events, corresponding to the fact that they will draw different conclusions as to whether the radiation they either observe or infer is real or virtual.

Suppose that L and L' both possess measuring instruments sufficiently sensitive to detect gravitational density waves. Furthermore, suppose that L' is freely falling in the field of M. According to the principle of equivalence L' can be treated locally as if it were a coordinate system accelerated by a nongravitational force. L will say that a body m, falling with L', radiates and also undergoes virtual momentum exchanges with surrounding bodies. L' will say that he and m exchange momentum through virtual skipping processes, but that m does not radiate except for the small accelerations resulting from the forces between them. But L' will nevertheless see radiation due to the fact that M is accelerating relative to him (even if M itself is not seen). Thus again there is no way of choosing a coordinate system so as to eliminate the linkage between virtual and real radiation. The topological structure of the vacuum density will be the same for L and L', but the actual densities reported will be different.

3.7.6. Comment on wave-particle duality. The connection between density waves and vacuum particle skipping should be distinguished from wave-particle duality. According to the vacuum particle interpretation the world line of a boson is a sequence of excitations of vacuum fermions accompanied by a compression-expansion wave in the fermion sea. Each vacuum fermion has a de Broglie wavelength, hence can be thought of as a matter wave. The compression-expansion wave can be thought of as a many particle process. These two dualities occur at different space-time scales. Nevertheless the occurrence of both wave-like and particle-like behavior is reminiscent of the photoelectric effect. When a skipping process is absorbed the density wave which accompanies it is annihilated as well. If the density wave carries real energy all this energy must be absorbed at the same time that the skipping energy is absorbed. Furthermore, the structure of the density waves must guide the skipping process, since the skips can only occur on those portions of the wave at which the skipping energy matches the vacuum density. The vacuum particle interpretation thus suggests that two separate levels of wave-particle duality operate in electromagnetic radiation and that the "needlelike" aspect of light is due to the strong correlation between skipping processes at the level of vacuum excitations and density waves involving the whole vacuum sea.

3.8. Mechanism of gravitation and Mach's principle. The analysis has assumed the depressing effect of mass and charge on the vacuum density implied by Eqs. (17) and (19). We have not, however, considered the mechanistic basis of this effect. This is key to the identification of mass with depression of vacuum density and for understanding how the mechanism of gravitation in the vacuum particle model connects to Mach's principle and the principle of equivalence. For the purposes of the present paper we must confine ourselves to how the theory accommodates these issues in principle. Formal proofs will be presented elsewhere.

3.8.1. Mechanism of mass induced depression. Positive energy mass can arise in two ways in the vacuum particle model. First, negative energy particles may be promoted to positive energy particles. In this case matter and antimatter are created. The production of a positive energy particle by this mechanism involves a transiently excitable particle from the vacuum, thereby locally reducing ρ_{grav}. This local reduction corresponds to the hole, or antiparticle. The promoted particle can cohabit the same fluctuation volume as a transiently excitable vacuum particle as long as this particle does not in fact undergo a transient excitation. As soon as it does undergo such an excitation, it will be repelled due to the exclusion principle, thereby again creating a local reduction in ρ_{grav}. Thus the appearance of two positive energy masses through pair production can be equated to the appearance of two depressions (or holes) in the vacuum density. In principle a depression could also appear through direct stretching of the vacuum. In this second mechanism positive energy matter would be created without the creation of antimatter.

Holes in the vacuum density (whether due to particle promotion or vacuum distension) act as absorbers of skipping processes. A region of high mass is a region of low vacuum density consisting of many such absorbing holes. If the pairwise inverse square gravitational interactions among transiently excitable vacuum particles are attractive, the particles will be attracted to regions of the universe with low positive energy mass, due to the fact that the density of gravitating vacuum particles is higher in these regions. As a consequence a vacuum hole acts as a repeller of vacuum particles. If a local hole or depression is created, the density of transiently excitable particles in the surrounding region will also be depressed. Actually this interaction is complicated by the fact that the gravitational force attracting an absorbing vacuum particle to a second one that is close to the massive body would become greater as the depressing effect develops, due to the $\rho_{grav}^{-\frac{4}{3}}$ dependence of the gravitational force (Eq. 16). A more important fact is that this dependence means that ρ_{grav} increases, therefore that ΔE_g is smaller. Consequently it is more likely that improper fluctuations of energy $\Delta E > \Delta E_g$ will occur, leading to a greater density of trapped skipping processes. This increases the likelihood that a vacuum particle in the neighborhood of positive energy mass will terminate skipping processes.

3.8.2. Mechanism of charge-induced depression. The situation with electric charge differs in an important respect. The creation of positive energy particles by either vacuum particle promotion or local dilation of the vacuum requires energy. The energy requirement is independent of whether or not the positive energy particle is charged. Furthermore, the sign of the charge does not change when a particle is promoted from a negative energy to a positive energy state. The positive electric charge associated with the resulting hole is not a new charge, but the absence of a charge. The important point is that negatively charged positive energy particles are affected by electromagnetic skipping processes in the same manner as charged vacuum particles in a state of trapped skipping. The only difference is that a charged particle becomes a continual initiator and terminator of electromagnetic skipping processes when it is promoted to the positive energy state. In the negative energy state it initiates and terminates propagative skipping processes only when it is undergoing trapped skipping.

Due to this continual emission and absorption a charged particle in the promoted state has a stronger repulsive effect on the surrounding charged particles in the vacuum than it does in its negative energy state. Hence a positive energy electron and a positron both polarize the surrounding vacuum sea of electromagnetic vacuum particles. The positive energy electron depresses ρ_{em}, in accordance with Eq. (19). The positron increases ρ_{em} since it represents a local deficit of charge, and therefore charged vacuum particles are repelled into its neighborhood.

3.8.3. Vacuum thermodynamics. Asymmetries of density tend to disappear in collections of long lived particles since more particles migrate from the higher density regions to the lower density regions than conversely. This, however, is not the case for vacuum particles, as shown by the following two arguments.

i. Energy argument. If the vacuum could homogenize as a result of virtual processes, the forces among masses and charges would disappear, hence all the potential energy would disappear without being converted to kinetic energy. Since mass is itself "vacuumized vacuum," homogenization of the vacuum would mean the disappearance of rest mass energy without the appearance of either potential or kinetic energy. Thus the stability of dilute regions of the vacuum is required by the principle of conservation of energy.

A corollary to this argument is that vacuum particles cannot homogenize by moving relative to one another without exchanging positions. That is, they cannot migrate into the "truly empty spaces" in between vacuum particles in dilute regions of the vacuum. These spaces may be viewed as potential barriers (and the vacuum particles as sitting in potential wells).

ii. Mechanistic argument. In order for a vacuum particle to migrate through the vacuum it must undergo a transient pair formation process (since it has zero velocity in its ground state) and a pathway for migration must also be available. This is because the transient pair must return to the ground state in a short amount of time. A hole must be available for the particle to fall into, and a particle must be available to fall into the hole. If no neighboring vacuum particle is simultaneously in a transiently excited state, the particle and hole must rejoin at the location at which they were originally created. Migration to the neighboring region is excluded by the Pauli principle unless a pathway of holes arising from transient pair creations is available. Since the Pauli principle only allows vacuum particles to exchange positions, no change in the vacuum density can occur (furthermore no migration is detectable since the particles are identical).

The mechanistic argument may be formulated in a more specific manner. The chance of a pathway for migration opening up in a more dense region of the vacuum is greater than the chance of one opening up in a less dense region. A transient pair in a more dense region is therefore less likely to move into a less dense region than conversely. Consider two adjacent regions, I and II. *The equilibrium rule is: the number of transient particles that are available for migrating from region I to region II is proportional to the density of region I, but the fraction of these which can actually succeed in entering region II is proportional to the density of region II.* As a consequence the entropy of the vacuum is conserved even if its density structure is inhomogeneous (entropy is here taken as a measure of the number of possible complexions of transient particle formations compatible with the density structure of the vacuum). By contrast, the entropy of a collection of positive energy particles increases because migration of particles from region I to region II depends only on the density of particles in region I, not in region II, and conversely.

The equilibrium rule for vacuum particles makes it clear why the density structure of the vacuum should be statistically stable. The structure is alterable only through the inducing effects of mass and charge, and through fluctuations, assuming conservation of energy. Particles, conceived as holes in the vacuum density structure, are also stable. No transient excitations can occur in such a hole. Therefore there is no pathway of migration into it, unless the transiently excited vacuum particle migrates into a hole which corresponds to its promoted antiparticle. In this case the only consequence will be that the hole will migrate. A particle which corresponds to a vacuum hole therefore appears inpenetrable to vacuum particles, while a particle which corresponds to a depression in the vacuum density would appear to have low penetrability. As a consequence both holes and depressions are stable structures. The migration of a hole also occurs when a promoted electron attracts a promoted positron, or when two positrons repel one another.

The frictionless motion of positive energy particles is due to the fact that they are in a different energy state than vacuum particles. Consequently the exclusion principle does not interfere with their occupying the same space. The only exception occurs when a transiently excited vacuum pair or hole is present at the same time as a positive energy particle or hole which is identical to it. In the case of an electron there are two possible consequences. If the positive energy electron has enough energy the collison can lead to the promotion of the virtual pair to a permanent positive energy status. Or the direction of motion of the electron may be altered. Such collisions are relatively infrequent since ρ_{em} is low. The problem may not arise with gravitational vacuum particles, despite high ρ_{grav}, because it is unlikely that they could occur as positive energy particles under normal circumstances due to their extremely high mass (Section 3.5.4).

The equilibrium rule is compatible with energy conservation, but would only entail it if permanent inhomogeneities are not allowed to arise through fluctuations. If this were possible, a homogeneous vacuum density would undergo spontaneous symmetry breaking.

3.8.4. Mechanism of gravitation. We are now in a position to give a proper description of the gravitational interaction between positive energy masses in the vacuum particle model. It is due to the fact that more skipping processes originate in regions of space with high vacuum density than in regions with low vacuum density. The vacuum density in the region of space between two positive energy particles will be more depressed than the vacuum density in the surrounding space. Each of these particles will in effect be bombarded by more skipping processes which push them together than push them apart. The virtual exchange mechanism responsible for gravity is thus an indirect one. The feature to note is that the repulsion of a vacuum hole is equivalent to the attraction of a vacuum particle. Thus a skipping process impinging on a vacuum hole pushes it away from the vacuum particle which is the source of that skipping process by attracting a vacuum particle in a state of trapped skipping into that hole.

The absorption processes, like the emission processes, directly involve vacuum particles in states of trapped skipping, not the "holes" or spaces between these vacuum particles. If these spaces could act as absorbers, skipping processes would in general be terminated in free space, in conflict with the conservation of momentum and in conflict with the existence of forces between positive energy bodies. However, it does not lead to any conflict if some absorption occurs in depressed regions of the vacuum since depression is equivalent to the presence of field. The probability is low, however, since the field falls off as an inverse square. More importantly, the feature which distinguishes a positive energy body from the depression it induces is a discontinuity in the vacuum density. Thus the probability of absorption changes discontinuously at the boundaries of a positive energy body. Propagative skipping can proceed in the presence of a continuously varying vacuum density, but if the vacuum density changes discontinuously, the skipping process must either be terminated by absorption or the velocity of light must change discontinuously.

We can now be more precise about the difference between the gravitational and electromagnetic interactions. While two positive energy masses attract each other because they are pushed together by the attractive interactions among negative energy particles in states of trapped skipping, any charges attracted to the positive energy masses repel each other in the same manner as charged negative energy particles. The direct effects are much more important than the induced effects in the case of charge, to some extent because ρ_{em} is much smaller than ρ_{grav}, and to a greater extent because the distribution of positive energy charge in the universe must on the average be neutral.

Finally we should note that it is possible to construct an alternative interpretation of gravitation in which all gravitational skipping processes are repulsive. The repulsive effect of positive energy mass on vacuum density would then be due to the fact that positive energy particles are permanent repellers, while the attractive gravitational interactions among these particles would be due to the asymmetrical effects of repulsive skipping processes acting directly on vacuum holes. This interpretation, however, has to deal with the problem of eliminating absorption by the spaces between vacuum particles.

3.8.5. Mechanism of inertia and principle of equivalence. The quantized inverse square law (Eq. 16) applies to the pairwise interactions of gravitational vacuum particles in states of trapped skipping. The indirect mechanism of gravitation between positive energy masses does not necessarily follow a simple inverse square law, though it may be shown that the interaction between two spherical masses is inverse square to a high degree of approximation. Rather than pursue such calculations here we will take a more conceptual approach, indicating how the vacuum particle interpretation leads to the principle of equivalence, hence to the principle of general covariance and to the gravitational law of general relativity.

According to the principle of equivalence the (gravitational) mass of a body as defined by the gravitational force can always be taken as equal to its (inertial) mass as defined by its resistance to acceleration. This is an expression of the fact that the acceleration of a body in a gravitational field is independent of its mass. As a consequence, it is impossible to distinguish (locally) between a gravitational field and an accelerating frame of reference. The principle of equivalence asserts that the behavior of a body in a gravitational field is locally equivalent to its behavior in some accelerated coordinate system [18].

Consider first a positive energy mass m, in an accelerated state of motion. The acceleration may be due to the presence of a second mass M, or it may have electromagnetic or other nongravitational origin. The mass m will produce a depression in the surrounding vacuum density. The vacuum particles in the immediate neighborhood of m are more likely to be in a state to absorb than vacuum particles in the increasingly dense surrounding space due to the fact that ΔE_g is smaller near m. As a consequence the change in the state of motion of the vacuum depression in response to the asymmetrical effects of gravitational skipping processes is retarded relative to the change in state of motion of m in a way that depends on distance from m. Because of this disequilibrium effect the vacuum density in the space into which m is moving is greater than it otherwise would be, increasing the relative number of gravitational skipping processes impinging on m that are opposite to its direction of motion. *Alternatively stated, the increase in the symmetry of bombardments and counterbombardments on a positive energy body resulting from the retarded motion of its induced vacuum depression provides a resistance to any change in the state of motion of that body.*

The vacuum particle model thus provides a natural mechanistic interpretation of inertia. Inertia is a form of gravitation. The model manifestly incorporates Mach's principle that inertial forces (or accelerations) are due to the distribution of (positive energy) mass in the universe. More precisely, they are determined by the distribution of "empty space," corresponding to the distribution of high density vacuum. The only difference between gravitational and inertial mass is that the change in state of motion of a body in the latter case can be initiated by forces of a nongravitational nature. As a consequence inertial mass is strictly a disequilibrium phenomenon arising from the motion of the body relative to the density structure of the vacuum, most significantly relative to the vacuum density depression that it induces in the surrounding space.

Now let us return to the equality of gravitational and inertial mass. If the initiating force can be turned off, the symmetry of bombardments and counterbombardments will become complete and the change in motion will come to a stop. This is not possible if the force is gravitational, since gravity is tied to the density structure of the vacuum. But for simplicity suppose that the gravitational field to which body m is subject is due almost exclusively to its proximity to M. As m increases, the depression of the vacuum density also increases. Since the biggest decrease occurs in the space between M and m, the number of bombardments per unit time pushing body m towards M increases relative to the number pushing it away. Thus the number of bombardments changing the state of motion increases as its mass increases. But this increase also serves to advance m relative to its induced depression, thereby increasing the intensity of counterbombardments opposing the change in state of motion. The acceleration of the body will thus be independent of its mass as long as the increase in the intensity of bombardment is proportional to the increase in the intensity of counterbombardment.

This argument provides a mechanistic interpretation of the equivalence principle, but it is not complete since the exactness of the proportionality is still open. Given the vacuum particle framework it is easier and more general to prove the principle directly with a thermodynamic argument. This is because inertial and gravitational mass are both controlled by the asymmetry of collisons with gravitational skipping processes, the only difference being that the asymmetry can be initiated by forces of a nongravitational nature in the measurement of inertial mass. Suppose, as the reductio hypothesis, that the inertial mass is not equal to the gravitational mass. If this were true it would mean that the ratio of gravitational to inertial mass of a body would depend on how far out of equilibrium it is with its surrounding vacuum density. But to disequilibrate the surrounding vacuum density it is only necessary to accelerate the body. The gravitational mass would, according to the reductio hypothesis, either increase or decrease. This would mean that the weight of the body would either increase or decrease when it is accelerated. By lifting a body with (without) acceleration and lowering it without (with) acceleration in the former (latter) case we could build a device in which the body weighed more on the way down than on the way up. Such a device could serve as a perpetual motion machine of the first kind, in contradiction to the principle of conservation of energy.

The red shift, the slowing of a clock in a gravitational field, the reduction of light velocity in the presence of mass, and the bending of light rays in a gravitational field all follow from the principle of equivalence [19]. The deduction of the principle of equivalence from the vacuum particle interpretation ensures that our indirect mechanism of gravity reproduces all these features. We could (but not here) take the analysis a step further. In the vacuum particle framework a body subject to inertial forces due to the acceleration of a coordinate system can always be interpreted as a body surrounded by a suitably chosen vacuum density structure. The principle of general relativity (that is, of the general covariance of the laws of physics) translates to the requirement that the equations describing the interactions among particles should be independent of the choice of vacuum densities. The vacuum density structure of the universe depends on the choice of coordinate system; but if it is possible for the observer to choose his state of motion he is free to choose it in such a way that the density structure is as homogeneous as possible. Following along this line we can see how the space curvature model of gravitation could be translated into a density structure model. The advantage of the vacuum density picture is that it provides for a common mechanistic basis for gravitation and electromagentism, yet allows gravitation to affect electromagentic phenomena and not conversely. This will be the situation as long as $\rho_{grav} \gg \rho_{em}$, which in turn is possible because these densities can vary independently due to the freedom afforded by the T_0^{em} term. If all forces could be interpreted in terms of space curvature it would be hard to isolate gravitational phenomena from electromagentic effects.

3.9. Summary of disequilibrium effects. We can now summarize the self-consistency requirements which lead to the equilibrium values of vacuum particle mass.

i. If vacuum particle mass is smaller than the skipping energy defined by the corresponding vacuum density, long skips will occur which increase the mass. If it is larger the force will be turned off altogether.

ii. On the average the gravitational mass of vacuum particles matches the local gravitational vacuum density, while the mass of electromagnetic vacuum particles matches the local electromagnetic vacuum density. As a consequence the electromagnetic vacuum particles carry kinetic energy in addition to the kinetic energy equivalent of their rest mass energy when they mediate gravitational interactions.

iii. The presence of a body with positive energy mass depresses the vacuum density surrounding it, and in fact the mass of the body can be thought of in terms of a depression of vacuum density relative to the average vacuum density of the universe. If the body carries charge the electromagnetic portion of the vacuum density will be further depressed.

iv. Local variations in the gravitational vacuum density are accompanied by variations in light velocity sufficient to maintain the equilibrium of mass as defined by gravitational force and mass as defined by skipping energy. Charge induced depressions have too small an effect on the total vacuum density to influence light velocity. In this case it is the alteration in the kinetic energy carried by electromagnetic vacuum particles mediating the gravitational interaction which maintains the equilibrium between vacuum particle mass as defined by gravitational force and as defined by electromagnetic vacuum density.

v. In both the gravitational and electromagnetic cases the occurrence of local variations in the vacuum density requires skipping processes to become enveloped by density waves which allow the skipping energy to remain constant from point of initiation to point of absorption.

vi. Inertia is a gravitational interaction that results from the disequilibration of a body and its surrounding density structure.

The equilibrium between vacuum particle density and vacuum particle mass is the result of averaging a large number of processes. The gravitational and electromagnetic fields flicker out of existence whenever the masses of the vacuum particles mediating these fields fluctuate above their threshold values. When the gravitational field flickers out of existence m_g^g falls back below the threshold, immediately turning the field on again. When the electromagnetic field flickers out of existence the decrease in the repulsions among vacuum particles leads to a compression of ρ_{em} which turns this field on again. These flickerings out of and back into existence are in general a highly local affair which has no macroscopic manifestations, just as the continual transition of atoms between two phases has no macroscopic manifestation. However, the question arises as to whether there might not be some conditions under which a phase transition could occur which leads to the disappearance of one of the fields over macroscopic regions of space.

For example, consider a highly condensed, massive star. The gravitational density within the star is much lower than that in free space and must exhibit a rapid spatial variation at least near the boundary of the star. The electromagnetic vacuum density will also be low and sharply varying near the boundaries if many of the promoted particles are charged. This is independent of whether the star has any net charge. As the star condenses there will also be a rapid temporal variation in ρ_{grav} and ρ_{em}. Recalling Eq. (89), the rapid decrease in ρ_{grav} pulls m_g^{em} up, whereas the rapid decrease in ρ_{em} pulls it down (since T_g^{em} increases when $\rho_{grav} - \rho_{em}$ increases). But since ρ_{grav} is the source of many more interactions than ρ_{em}, it is unreasonable to expect that the rates of the two equilibration processes are equal. The rate at which m_g^{em} is pulled up by the rapid decrease in ρ_{grav} should be faster than the rate at which it is pulled down by the rapid decreases in ρ_{em}. When the disequilibrium exceeds a critical point the minimum energy required for pair production will become too high for electromagnetic skipping to occur.

Since all electromagnetic vacuum particles in the region of rapid density variation will exceed this threshold point at about the same time, the electromagnetic field will be turned off in a macroscopic spatial domain. The only force which will remain in this domain is the attractive force of gravitation (ignoring weak forces and nuclear forces, which are immaterial here, but which in any case can be interpreted according to broken symmetries in the vacuum density structure). The vacuum particles in this repulsion free domain will thus condense and in addition attract vacuum particles from surrounding regions of space. As a result the vacuum density will increase, allowing the energy required for electromagnetic pair production to again be less than or equal to the energy required for electromagnetic skipping. The electromagnetic interactions will reappear on a macroscopic scale, switching on the repulsive interactions among the electromagnetic vacuum particles and consequently again lowering the vacuum densities, and so on. This cyclic switching on and off of the electromagnetic field will evidently be accompanied by periodic coherent accelerations of charged particles and therefore with the emission of periodic bursts of radiation into the universe. Such instabilities in the vacuum structure may be the origin of various forms of cosmic radiation, such as the periodic bursts of radiation emitted by pulsars.

4. Measurement and Irreversibility

The vacuum particle interpretation is a mechanistic model of the vacuum constructed to fulfill the requirements of quantized force laws. While it uses conceptual devices which have their origin in quantum mechanics it is clearly not a quantum mechanical theory in the usual sense. The status of the model may be clarified by assuming that forces are mediated by vacuum particle skipping and using this picture to repeat the initial derivation of the quantized force laws in a more defined manner [20].

Let $\hat{p}(i \rightarrow f)$ denote the probability that a vacuum particle initially in the highest negative energy state is promoted to a positive energy state in which the skipping energy corresponds to the vacuum density. This local transition probability in first order is given by

$$\hat{p}(i \rightarrow f) = \frac{4|<f|V|i>|^2 \sin^2\left\{\frac{1}{2}(\Delta E(\tau_K, x_K)/\hbar)\tau_K\right\}}{\Delta E(\tau_K, x_K)^2} \qquad (99)$$

The skipping energy is given by $\Delta E(\tau_K, x_K) \approx \hbar/\tau_K$. Because of the requirement for macroscopic energy and momentum conservation in propagative skipping, this is the only energy shift which can contribute to a force. Furthermore, τ_K is the only time interval that is relevant, for processes occurring during shorter times either do not contribute to propagative skipping or would redefine the skipping energy, while skipping processes occurring over longer times are excluded since the skipping energy is too low for pair production in this case. Substituting

$\Delta E(\tau_K, x_K) \approx \hbar/\tau_K$ into Eq. (99)

$$\hat{p}(i \rightarrow f) \approx \frac{0.92| < f|V|i > |^2 \tau_K^2}{\hbar^2} \qquad (100)$$

This step appears to differ from the initial derivation in that no use is made of the linearity assumption (that the magnitude of the fluctuation required to achieve a given degree of momentum transfer is linearly dependent on the distance). But it should be remembered that Eq. (100) refers to a local skip, not to the total exchange between two absorbers. The restriction of the skipping energy to $\hbar/\tau_K$ is a consequence of the linearity assumption, which is therefore no longer needed once the skipping mechanism is admitted as the starting point of the analysis.

Now consider two absorbing particles separated by a distance r and denote the probability that a skipping process initiated in the neighborhood of one is absorbed in the neighborhood of the other by $p(i \rightarrow f)$. Between the two particles under consideration there are $N = r/c\tau_K$ steps of the skipping process. The absorbing particle is located on the surface of a sphere of area $4\pi r^2 = 4\pi N^2 c^2 \tau_K^2$. If the vacuum density is homogeneous each fluctuation domain on this sphere has an area of approximately $\pi c^2 \tau_K^2/2$. Thus the number of possible positions of the absorber is approximately $8N^2$ and the chance of its actually being absorbed is approximately $1/8N^2$. Since macroscopic energy conservation requires that an initiated skipping process be propagated until it is absorbed

$$p(i \rightarrow f) \approx \frac{\hat{p}(i \rightarrow f)}{8N^2} \approx \frac{0.12| < f|V|i > |^2 \tau_K^2}{\hbar^2 N^2} \qquad (101)$$

Taking the force between the two particles as proportional to this probability

$$\mathbf{F} \approx \frac{A| < f|V|i > |^2 \tau_K^2}{\hbar^2 N^2} \frac{\mathbf{r}}{r} \qquad (102)$$

which is the quantized force law in terms of fluctuation times.

Now consider the matrix element $< f|V|i > = \int \psi_f^* V_{op} \psi_i \, d^n q$. As an approximation ψ_i may be identified with the wave function of the free particle in its highest negative energy state. ψ_f may be identified with the wave function of the particle-antiparticle pair in a positive energy state corresponding to the skipping energy. In the case of electromagnetic vacuum particles, ψ_i and ψ_f may be tentatively taken as given by the Dirac equation, with ψ_f corresponding to the lowest positive energy state for skipping processes mediating the electromagnetic field. The matrix element can be interpreted in analogy to a Feynman diagram, with ψ_i representing the probability amplitude that the vacuum electron has coordinates $q_1, \ldots, q_n$ in state i, V_{op} representing the amplitude that the perturbation potential acts on the particle, and ψ_f^* representing the amplitude that the particle leaves the point with these coordinates in the excited state. The perturbation potential may have any value, but for initiating a propagative chain it should be taken as the skipping energy itself, that is, as $\hbar/\tau_K$. This may be interpreted as a local fluctuation in energy required by the uncertainty principle, but compatible with formal conservation of energy in a larger system, or as a true violation of conservation of energy in the sense of being a deviation from translational symmetry in time. The latter interpretation is to be preferred, however, since in the vacuum particle interpretation such deviations are the source of all interactions.

The reason for viewing ψ_i and ψ_f as conceptual approximations is that in the vacuum particle model mass arises from a bootstrapping process that is more global than the renormalization process of standard quantum field theory. Admitting ψ_i and ψ_f as approximate descriptions is thus reasonable only when the whole set of mass-vacuum density relations has reached equilibrium, at which stage a local description in terms of an equation of motion becomes admissible.

At this stage the quantized force law serves only to provide a slightly modified form of the potential function through the relation $\mathbf{F} = -\nabla U$ (along with the substitution $N = r/c\tau_K$). Whether the introduction into the potential function of fluctuation quantities such as τ_K actually leads to quantitative agreement with, say the Lamb shift, without requiring any arbitrary calculational procedures is a question which obviously requires investigation. But it might be noted that the quantized potential functions reduce to the classical ones when τ_K is taken as one unit of time in the appropriate dimensions independent of r (or, alternatively, if r_K or ρ_K are taken as one unit of length or density in the appropriate dimensions). From the point of view of the vacuum particle model this is a pathological assumption which is buried in the choice of constants in the conventional theory.

In principle a conceptually proper description should begin with the whole vacuum structure. The initial wave function in the matrix element should represent the amplitude that the absorbing particle and vacuum are in state i, V_{op} should represent the amplitude that the neighboring vacuum particle undergoes an uncertainty fluctuation, and the final wave function should represent the amplitude that the absorbing particle and vacuum undergo a transition to state f. Such a global description is definitely necessary whenever the equilibrium relations between mass and vacuum density are changing in a significant way. Quantized force laws provide a natural mechanism for incorporating such a global description. For this purpose denote the wave function ψ_i by $\psi_i(q_j, j, t)$, where q_j represents coordinates associated with particle j and t represents time. These are still regarded as free fermion wave functions, but initially unknown and not provided by the solution of any equation of motion which refers to a single particle. We suppose that the wave function of the vacuum as a whole can be written as

$$\Psi_K(q_1, 1; \dots ; q_j, j; \dots ; q_n, n; t) = \det_{j \in K} \psi_i(q_j, j, t) \qquad (103)$$

where the notation symbolizes that this should be a Slater determinant with j running over the gravitational vacuum particle when $\psi_K = \psi_{\text{grav}}$ and over all vacuum electrons if $\psi_K = \psi_{\text{em}}$. Assuming that position coordinates are sufficient to characterize the vacuum density we can use the usual interpretive postulate of quantum mechanics to write

$$\rho_K = |\Psi_K(q, t)|^2 \qquad (104)$$

where $\Psi_K(q, t)$ is shorthand for the left hand side of Eq. (103). The quantized force law can now be expressed as

$$\mathbf{F} \approx \frac{A |<f|V|i>|^2}{c^2 \hbar^2 \{|\Psi_K(q, t)|^2\}^{\frac{1}{3}} r^2} \frac{\mathbf{r}}{r} \qquad (105)$$

The new and important feature introduced here is that the square law form $|\Psi_K(q, t)|^2$ represents the results of an irreversible projection (or measurement) process which is ordinarily distinct from the time evolution process. The time symmetry of any system governed by a potential function which incorporates such a square law must be broken. When the relations between vacuum particle mass and vacuum density have reached equilibrium and the vacuum structure is essentially static $\Psi_K(q, t)$ can be replaced by $\Psi_K(q)$. Such a regime can be reversible and can be locally described by reversible time evolution equations (such as the Dirac equation) since all the dissipation necessary for the vacuum structure to reach its given state of organization occurred in the past. There is no way of detecting that a square law is operative in such a regime since it has already been converted to a number, the probability of some vacuum structure. But when the vacuum structure undergoes significant enough change for the relations of mass and vacuum density to depart from equilibrium, $\Psi_K(q,t)$ and the $\psi_i(q,j,t)$ must both evolve until a self-consistent equilibrium is reached. It is in this bootstrapping regime that the operations of the square law become manifest and the time symmetry is broken.

Conceivably the projection process could be integrated with the time evolution process by constructing a potential function which incorporates $\Psi_K(q,t)$ rather than $|\Psi_K(q,t)|^2$. But a complex potential represents an absorption process and therefore would not alter the situation as regards irreversibility [8]. The chief point, conceptually, is that the vacuum particle model implies an interaction between the micro- and macrophysical structure of the universe that is analogous to that between a microphysical system and the macroscopic constraints of a measuring instrument. Early in the history of the universe, before an equilibrium between mass, charge, and vacuum density was established, wave function collapse would have been a dominant feature. After equilibrium was established this feature would have become masked in the small and in general lost in ergodic type entropy increase in the large. It may have remained significant for some astrophysical phenomena, such as the pulsar effects suggested. The origin of biological systems and measurement instrumentation are another regime in which the underlying wave function collapse would be unmasked, in this case due to sensitive nonlinearities that amplify the collapse effect to macroscopic actions.

Measurement instrumentation that implements square law type projection operations are an extreme case, since here collapse can be complete and sudden, with the time evolution of the wave function as captured in equations of motion completely suppressed. This is probably an idealization of the real situation in all practical cases. Nevertheless, even if collapse occurs in stages and mixes with motion in the nonequilibrium universe, the question arises as to whether each phase randomization event occurs in a sudden manner applicable to all observers in the universe. If so, this would clearly be a violation of special relativity, just as in ordinary quantum mechanics and quantum field theory. If collapse is spontaneous and becomes more prominent as the region of space-time considered becomes larger, it would provide the intrinsic asymmetry required of a time coordinate.

5. Concluding Remarks

As stated more than once, we are dealing with a first step model. The modeling methodology it points to calls for recursive self-consistency. In the standard quantum mechanical methodology the analysis begins with a classically picturable description, prepares a nonclassical description for the purposes of calculation, and then collapses this back to a classical description for the purpose of relating the calculation to measured values. This is not in general possible in the vacuum particle model, since the forces required for the initial classical description are themselves determined by the evolving relationship between the structure of the vacuum sea and the visible distribution of mass and charge.

The whole picture is highly evolutionary. The microstructure of the universe interacts with the macrostructure, undergoes random "mutations" in phase, the macrostructure changes, and the process continues until self-consistency is reached. The situation is very similar to that of an evolutionary ecosystem, with the biotic component undergoing genetic mutations, the macroscopic constraints of the environment undergoing change, and the whole process continuing until self-consistency is achieved. Biological evolution, rather than being an awkward add-on to a dead universe, becomes a special case, albeit one in which highly correlated constraints are operative [21].

The situation also fits to the hypothesis that the evolutionary process of variation and selection allows biological nature to recruit unpicturable microphysical processes within cells for the integrated behavior of organisms and for cognition and intelligence in the context of the macrophysical architecture of the brain [22]. Followed to its evolutionarily opportunistic conclusion, this leads to the same kind of recursive interaction between the vacuum structure and the distribution of positive energy particles that in the vacuum particle model underlies the evolution of the universe in the large, and that unmasks wave function collapse and irreversibility in the special case of biological evolution and in the even more special case of laboratory measurement.

Acknowledgment. This research was supported in part by National Science Foundation Grant IRI87-02600.

References

1. von Neumann, J., *Mathematical Foundations of Quantum Mechanics,* Princeton University Press, Princeton, N.J., 1955.

2. Wigner, E. P., *Symmetries and Reflections,* Indiana University Press, Bloomington, Indiana, 1967.

3. Pattee, H. H., "Physical Problems of Decision-Making Constraints," in *Physical Principles of Neuronal and Organismic Behavior*, M. Conrad and M. Magar, eds., Gordon and Breach, New York, 1973.

4. Josephson, B. D., "The Artificial Intelligence/Psychology Approach to the Study of the Brain and Nervous System," in *Physics and Mathematics of the Nervous System*, M. Conrad, W. Güttinger, and M. Dal Cin, eds., Springer-Verlag, Heidelberg, 1974.

5. Bell, J. S., "On the Problem of Hidden Variables in Quantum Mechanics," *Rev. Mod. Phys.*, 38, 447.

6. Dirac, P. A. M., *The Principles of Quantum Mechanics*, 4th ed., Oxford University Press, 1958.

7. Roman, P., *Introduction to Quantum Field Theory*, Wiley, New York, 1969.

8. Schiff, L. I., *Quantum Mechanics*, 2d ed., McGraw-Hill, New York, 1955.

9. Bohm, D., *Quantum Theory*, Prentice-Hall, Englewood Cliffs, N.J., 1951.

10. Terrell, J., "Invisibility of the Lorentz Contraction," *Phys. Rev.*, 116 (1959), 1041-1045.

11. Weisskopf, V. F., "The Visual Appearance of Rapidly Moving Objects," *Phys. Today*, 9 (1960), 24-27.

12. Penrose, R., "The Apparent Shape of a Relativistically Moving Sphere," *Proc. Camb. Phil. Soc.*, 35 (1969), 137–139.

13. Rosser, W. G. V., *The Special Theory of Relativity*, Butterworth and Co., London, 1964.

14. Frisch, D. and Wilets, L., "Development of the Maxwell-Lorentz Equations from Special Relativity and Gauss's Law," *American J. Phys.*, 24 (1969), 574-579.

15. Tessman, J., "Maxwell–Out of Newton, Coulomb, and Einstein," *American J. Phys.*, 34 (1966), 1048-1055.

16. Weinberg, S., "Perturbative Calculations of Symmetry Breaking," *Phys. Rev. D*, 7 (1973), 2887-2910.

17. Condon, E. U., "Basic Electromagnetic Phenomena," in *Handbook of Physics*, E.U. Condon and H. Odishaw, eds., McGraw-Hill, New York, 1967.

18. Bergmann, P. G., *Introduction to the Theory of Relativity*, Prentice-Hall, Englewood Cliffs, N.J., 1942.

19. Einstein, A., "The Foundation of the General Theory of Relativity," in *The Principle of Relativity,* by H. A. Lorentz, A. Einstein, H. Minkowski and H. Weyl, Dover Publications, New York, 1923.

20. Conrad, M., "Reversibility in the Light of Evolution," *Mondes en Developement,* 14 (1986), 111-121.

21. Conrad, M., *Adaptability,* Plenum Press, New York, 1983.

22. Conrad, M., "Microscopic-macroscopic Interface in Biological Information Processing," *BioSystems,* 16 (1984), 345-363.

Patterns of Evolution and Patterns of Explanation in Economic Theory

GERALD SILVERBERG

Abstract

The scientific enterprise can be regarded as an exercise in pattern recognition in theoretically defined abstract spaces. From this point of view, economics is still underdeveloped in terms of agreement about what objects (regularities) to look for and about which basic principles will lead to the identification of appropriate frameworks for analysis. Various approaches to individual behavior, interaction and system-level emergent properties are contrasted in terms of their use of such explanatory principles as competition, equilibrium, search and rationality. A case is made against the notion of economic equilibrium and in favor of temporal and structural regularities within populations characterized by irreducible but endogenous diversity and subject to continual evolutionary transformation.

1. Introduction

Economics occupies a unique position between the natural or "hard" and the social or "soft" sciences. On the one hand it can look back on a long tradition of continuous theoretical development and increasing mathematical sophistication now almost rivaling mathematical physics. On the other, basic conceptual issues are still subjects of considerable controversy, while it remains unclear what if any real, empirically sustainable generalizations have actually been established. As Kalecki is reported to have put it some years ago, economics consists of theories which cannot be applied and empirical regularities without any theoretical basis. This state of affairs, if the presidential addresses to the American Economic Association of a number of distinguished economists, not to speak of the numerous "heretics" in the profession, are to be believed, has only worsened since then.

In this paper I would like to step back from strictly technical issues which have dominated the discussion in the past to examine a few central conceptual problems which in my opinion are at the root of economic science's present difficulties. To do so it will first prove necessary to outline a conceptual framework broad enough to encompass the most important competing approaches and evaluate their relative merits and mutual consistency. Then I will attempt to bring to bear some results deriving from mathematical biology, dynamic systems the-

ory and formal theories of evolution and learning, many of which may
be new to economists, to shed some new light on a few methodological
questions which, although they were formulated quite unambiguously
a number of years ago, continue to hover unresolved over the profession
like an unexorcized ghost. In the process I will also make use of some
more concrete models of my own to illustrate some of the main points.

2. Patterns of Phenomena and Patterns of Explanation

The natural sciences have been characterized by such a fruitful interplay of observation, formalism, theory, tests and new, often totally
unimaginable observations and applications that it has hardly ever been
necessary to ask what constitutes a scientific explanation and what are
the sort of objects to be explained. The process of scientific advance,
by unperturbedly throwing open its own track as it went along, made
these questions practically superfluous, relegating them to a philosophical rearguard action struggling to catch up in the dusty wake of the
main effort. The same cannot be claimed of economics. To begin
with, there does not even seem to be agreement about what exactly
the phenomena and issues actually are with which research must deal.
For example, the exclusively "equilibrium" orientation of the dominant
neoclassical school addresses the question originally raised by Adam
Smith concerning the coherence and welfare of an economic system
composed of individuals blindly pursuing their self-interest. Stable full
employment is seen as the "natural" reference state, so that unemployment over shorter or longer periods of time must be due to exogenous
and irreducible "shocks" or perverse policy interventions (depending on
the political point of view, by either government, unions or the central
bank). Other economists have regarded unemployment as the "natural" feature of capitalist economies demanding explanation, which has
lead to a multitude of business cycle and crisis theories and the theory of effective demand. It is apparent that what is considered to be
the object of analysis and the methodological perspective are somehow
inextricably intertwined. Moreover, ideological biases inevitably cloud
what one would like to think are strictly scientific issues.

Of course, these points have been debated at length during the
whole history of economic analysis. To avoid the sterility of these
controversies, I will try to abstract from the political/normative issues
(itself a dubious ideological undertaking, some would argue) and focus
instead on the kind of phenomenological regularities which have in
fact been uncovered and which could serve as an avenue of attack for
creating an appropriate dialogue between theory and empiricism.

One could make a case for regarding the scientific enterprise as an exercise in pattern recognition in geometric spaces inaccessible to unaided human perception. From this point of view, it is paramount to identify the variables which constitute those spaces in which "structurally stable" objects will in some sense stand out. Let me illustrate what I have in mind here.

Technical change is clearly one of the main springs of economic change. The traditional static factor allocation approach to production looked at it as a response to changing factor prices, a kind of substitution process appended to the posited substitutability of capital, labor, and land, and basically a passive response. This gives no indication, however, how the process evolves over time except to suggest that it is may be an extremely rapid and costless adjustment to exogenous noneconomic variables. If instead we regard technologies as something concreter than abstract relations between inputs and outputs, for example as distinct and enumerable types of machines or technology "species," then indeed a pattern does emerge.

Figure 1 summarizes data first collected by Fisher and Pry 1971 on market or production capacity shares of competing technologies. This represents a kind of census of machines, if you will, and makes no assumptions about underlying decision processes and profitabilities. It clearly shows that we are dealing with an approximately logistic substitution process extending over considerable periods of time. This undisputed and remarkably invariant phenomenon has been subsequently approached from two angles. Economists (such as Griliches 1957, David 1975, Davies 1979 and Mansfield 1968) have attempted to integrate it with traditional economic categories, particularly with microeconomic decision processes. The other approach has recourse to a higher level perspective and regards technological diffusion as a niche filling process obeying its own intrinsic, and, compared with the fickleness of economic predictions, robust laws. Fisher and Pry seem to have initiated this tradition, which has been carried forward in particular by Marchetti and Nakicenovic (representative is e.g. Marchetti and Nakicenovic 1979).

In the search for regularities let me briefly mention two further categories. The first is the search for temporal regularities, which has a long history and, as some observers have suggested, itself is subject to recurrent intellectual cycles. Beginning with the nonstatistical, impressionistic observations of the nineteenth century (or should we refer to the Biblical seven fat and seven lean years?), to Jevons' correlation of economic activity with the sunspot cycle, to the first moving average

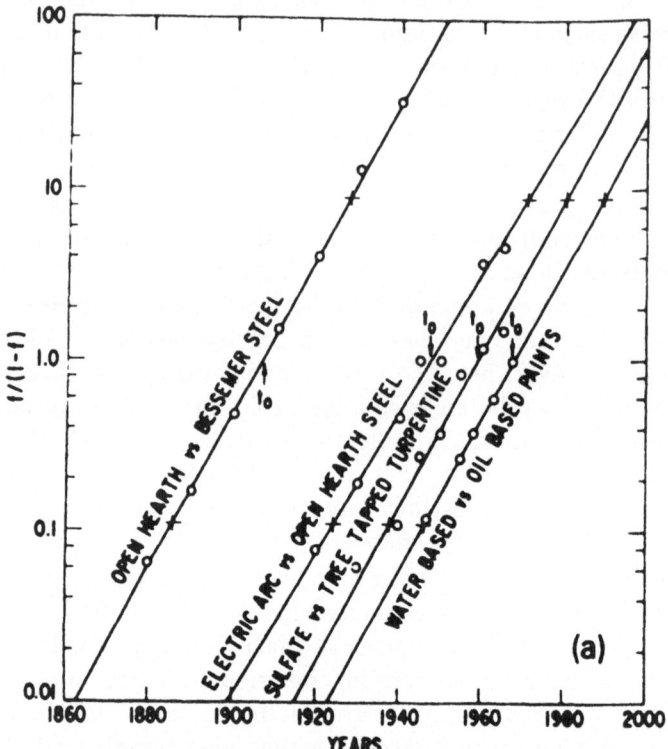

Figure 1. Diffusion Curves for Different Technologies
(after Fisher and Pry, 1971)

time series analyses, up to contemporary spectral analysis, chaotic dynamics and entropy dimensions, cases have been made for the existence of regular cycles of a range of periods from four-year inventory cycles to 50-year Kondratieff waves. This has lead from syntheses calling for a superposition of discrete spectra (such as Schumpeter 1939) to chaotic attractors, where a continuous spectrum of cycles of all periods occurs due to a stochastic or a nonlinear, low-dimensional deterministic mechanism. A whole literature exists almost independent of the empirical debate dealing with economic explanations for temporal regularities in industrial, regional, national and world economy settings.

The other category includes a number of population or frequency distributions of relatively invariant form, such as income distribution, the size of firms, or the distribution of stock market price changes (on a daily, weekly, monthly etc. basis). These, too, seem to fall into regular patterns, such as log normal or Pareto in the tails. There is a considerable literature addressing these questions, and a number of mechanisms have been advanced to explain them, a subject I will return to below. These phenomenon are not merely curiosities, however, but rather sometimes tie into such basic issues as the efficient markets hypothesis. In any event they offer a scientific access route into a subject that is fraught with the danger of ideological leaps of faith or overly axiomatic ambitions which never find their way back to reality.

3. Are There Consistent and Plausible Explanatory Foundations for Economic Theory?

One answer to the provocative question with which this section begins is that the majority of economists certainly (although perhaps only implicitly) believe that there are, and furthermore that they are already known and can be systematically applied. I will attempt a radical simplification at this point by reducing the main attempts at fundamental economic "explanation" to three formative principles: the rationality postulate, some concept of equilibrium, and competition. It is naively assumed that the first two principles, which are very tightly interrelated, receive their ultimate justification from the third. As I shall argue in the following, however, not only is this probably not true, but the first two concepts, at least as they are commonly understood, may not even be always compatible with the principle of competition when this is understood in its underlying evolutionary sense. To do this I will first make something of a detour to set out a skeleton structure for analyzing these questions.

4. A Recipe for a Generic Economic System

Our generic economic system will consist of a two-level hierarchy of basic units (e.g. households and firms) in interaction with one another such that an aggregative feedback dynamic emerges (the "market," the "economy"). The flows in Figure 2 may represent products, services, financial instruments, as well as informational variables such as orders, prices, and delivery delays communicated between units. Two "exogenous" factors have also been introduced: "novelty" and the "environment." Novelty denotes those micro changes which cannot be captured in this kind of structural diagram: the insertion at an historical point of time of hitherto untried strategies or tastes, technologies, new arrows of interaction or the removal of old ones, producing the redefinition of important subgraphs. The environment consists of everything which could influence the system but, at least to a first approximation, is

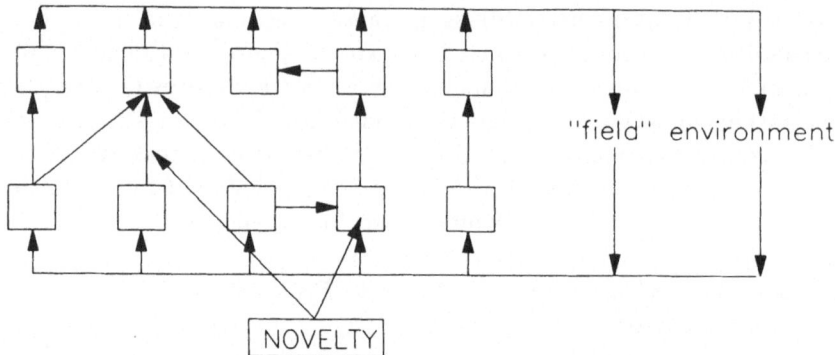

Figure 2. Elements of a Generic Description

not itself significantly influenced by it (such as the weather and the distribution of mineral resources).

To understand the interactions which can take place between the boxes it is necessary to outfit them with generic features which can then be adapted to their specific roles. At the input/output level we can identify information, goods, services and money as arrows entering and leaving the boxes. Four main categories can be differentiated as internal state variables.

(1) Unchanging characteristics of agents. These can be physical characteristics such as height and weight, genetic endowment, IQ. Economic theory has traditionally regarded tastes and technology as belonging to this category. It is obvious that the time frame of reference is decisive here, and that what is unchanging and what is variable in a given setting may be more a matter for empirical verification than a priori choice.

(2) Endogenously changing characteristics, such as stock variables deriving from the inputs and outputs: financial balances, productive capital, but also productivity, prices, knowledge and skills (socalled human capital), etc.

(3) Control or decision variables which, within certain bounds, can be consciously set by the agents to steer their interactions with the outside world. Examples are the level of discretionary spending, portfolio choices, rules of thumb such as payback periods and markups, choices of technique, etc.

(4) As Rosen 1985 has recently reemphasized, biological and social systems are characterized by a unique emergent property: they are anticipatory. This means in our case that at some conscious

or unconscious level decisions are taken on the basis of a model of the system's environment and its future evolution, either as an autonomous system (what the economist would call a game against nature) or as a function of the agent's own behavior (and thus as some sort of multiperson game). As we shall see, this is one of the major points of unresolved tangency between systems theory and orthodox economics, which have taken distinctly different methodological positions with respect to this problem.

We can now combine these ingredients according to Fig. 2 into some kind of dynamical model by introducing an explicit system of time dependence and recursions transmitted by the arrows and the internal dynamics of the boxes. Depending on whether we choose a deterministic or an explicitly stochastic formulation, this will result in a system of difference or differential equations of very high dimension, or even an integro-differential system. Stochastically it will lead to a master or Fokker-Planck equation formulation. The possible evolutions of such a system are inconceivably complex once one leaves the linear regime. Furthermore, the question of how to incorporate novelty is one of the most difficult aspects. One way around this problem, which I will discuss in more detail later, is to allow for enlargement of the state space or for perturbations in the structural equations and investigate the structural stability of the resulting system to see which novelties will in fact trigger qualitative change.

How does one set about analyzing such a system? The obvious questions is whether this complicated state space can be effectively reduced to one of much smaller dimension centered around some kind of privileged structure of simpler topological type which suffices to characterize the dynamic behavior. This endeavor, interestingly enough, is common to both modern systems dynamics and the standard tool of equilibrium analysis in economics (although in the latter, only recently have dynamic questions been explicitly addressed), even though, as I shall now argue, the methodological justifications are radically different.

5. The Search for Privileged States and the Rationale for their Privileged Status

In dynamic systems theory the standard method of analysis is to look for limit sets of the flow, and in particular for attractors. It is clear, at least for dissipative systems, that this will lead to a significant reduction in the relevant degrees of freedom of the system. Classically, one was interested in the existence and stability of point attractors or equilibria, for which an exhaustive classification using linear stability analysis is possible. The attractor zoology expanded in Poincaré's day to limit cycles and tori, while nowadays the hottest topic is chaos and strange attractors of fractal dimension. The rationale for this approach is intuitively plausible. After a sufficiently long period of time has elapsed (the relaxation time) the behavior of the system will settle down to one of the attractors and stay there for all time. Everything before this is merely a transient, and is treated as such (i.e., disregarded, though of course one can and often does study transient behavior for its own sake). The motion of the system itself, as either a transient or a steady state, is causally "explained" by its equations of motion, with causality running from the past to the future even if anticipatory mechanisms are involved. Simple dynamical models of this sort, both with and without expectational mechanisms, have been common in economics since the 1930s, almost exclusively in the macroeconomic domain, i.e., without an explicit foundation in the decision processes of the underlying economic agents and in their interaction and aggregation. Examples are the linear multiplier-accelerator business cycles, the Harrod-Domar growth model, and the various nonlinear cycle and/or growth models that began to appear in the 1940s (Kaldor business cycle, Hicks ceiling/floor cycle, Goodwin nonlinear accelerator and Lotka-Volterra growth cycle). Characteristic of this sort of modeling is the translation of impressionistic observations of aggregate behavior or so-called stylized facts into simple mathematical structures without invoking any general postulates about the ultimate foundations of economic explanation.

The "first principles" approach in economics, in contrast, is based on the one hand on a conception of individual behavior known as rationality and optimization, and on the other on a state of mutual consistency of these strategies known as economic equilibrium. It must be emphasized that the definition of economic equilibrium in either its general equilibrium guise (à la Arrow/Debreu in the tradition of Walras) or in the multiple game theoretic senses it has been given is quite different from its homonym in dynamic systems. As we shall see, economic equilibrium is a self-consistency condition imposed on the space of all possible behaviors under the assumption of "rationality." It is often suggested that once the system is in equilibrium (and additionally all agents know that this is so), no incentive exists for agents to change their behavior, and thus this state would be persistent. Even if this were the case, however, this notion of equilibrium is still a long way from the status of a (local or global) attractor in some suitable dynamic (and by this I mean temporally out of equilibrium) generalization of the usually static models.

What, then, is the peculiarly economic meaning of equilibrium, both as a mathematical tool to focus on a "privileged" subset of the state space and as a causal justification for this reduction in dimensionality. First, common to all "neoclassical" approaches is a specific choice-theoretic description of the behavior of the boxes in Figure 2. A known relationship is posited between a well-defined and comprehensive set of actions (the choice set) and a set of outcomes, the latter admitting a cardinal valuation in terms of von Neumann-Morgenstern utilities. This formulation of utilities readily generalizes to situations of actuarial uncertainty by going to expected utilities over either known probability distributions over states of the world or subjective distributions subject to Bayesian learning. The action actually taken will be the one which globally maximizes (expected) utility, instantly and costly identified as a result of rational, conscious calculation and not of some process of experimentation, gradual learning, social routine or the like. Whether this paradigm of individual behavior is fruitful or valid is still a highly controversial question in itself (significantly, psychologists in the main do not seem to share it, and there is considerable direct experimental evidence which appears to contradict its basic axioms; cf. for example Tversky and Kahneman 1986). The problem for all social theories, however, is that the boxes in Figure 2 are not alone in a game against nature but are engaged in a complex matrix of dynamic interaction with each other.

The general equilibrium approach employs a simplifying assumption to reduce this tangle of interdependent choice to a self-consistent game against nature after all: the concept of parametric free competition. If the agents are so numerous and small that their own actions are too insignificant by themselves to reasonably affect their decisional environment, then the momentary data of the economy can be taken as fixed parameters in their decision calculation, which reduces to a maximization problem with respect to their control variables.

The essence of general equilibrium theory is an existence theorem which, by invoking a fixed point theorem, demonstrates that under certain assumptions about preferences and production, there exists a set of prices making the individual decision problem and the parameters entering into it (resulting from aggregation) mutually consistent. What the economy would do in real time were the system still to be outside "equilibrium" (i.e. for prices at which the quantities demanded and produced as a result of individual parametric decisionmaking were not in balance) is left totally unspecified. In other words, this notion of equilibrium does not automatically induce a flow on the state space, nor is it necessarily an attractor of any particular plausible flow one might choose to employ (the stability theorems of general equilibrium theory which appear to address this question are only valid under even more restrictive assumptions and do not purport to describe realistically "out of equilibrium" decisionmaking). Instead, an off-line coordination process is often invoked known as the tatônnement auction in which iterative adjustments are carried out before any actual economic trading or production takes place.

As I shall argue below in the context of the dynamics of collective phenomena, it is not at all clear that the perfect competition simplification (apart from whether it is a realistic characteristic of modern corporate capitalism) can serve to reduce system complexity in this way in the face of interdependent but "boundedly rational," imitational behavior, even when the agents are infinitesimally small. One need only think of mobs, fads and fashions, and, as Keynes argued (Keynes 1936, p. 156–157), the stock market to understand this point.

Within the framework of "rational choice behaviorism" the most direct attempt to confront the irreducible problem of social interaction is of course game theory. The choice environment is no longer seen as given, but rather as itself the result of actions of other agents. Agents still dispose over well-defined choice sets, but utility outcomes are no longer single-valued functions of these choices but columns of a multidimensional matrix reflecting the choices of all other agents. The difficult question once again is to define some privileged subset of all possible behavioral configurations which for some compelling reason should be the focus of analysis. The solution once again is some notion of equilibrium, i.e., some configuration of behaviors such that, should agents know that they are in it and understand the game they are in "properly," i.e. "rationally," they would have no incentive to unilaterly deviate from.

Unfortunately, there is no unique definition of the equilibrium of a game which naturally satisfies all these intuitive requirements once the zero-sum setting exhaustively analyzed by von Neumann and Morgenstern 1944 is left. The most widely used notion is that of Nash which, for given behaviors of the other players, calls for player i to maximize his payoff. This is then symmetrized for all players, i.e., an analogous form of mutual consistency must be imposed as in general equilibrium theory. Not surprisingly, Nash's existence theorem for equilibria (albeit in games with finite strategy spaces) has recourse to Brouwer's fixed point theorem. However, the notion of "rationality," which seemed so unambiguous in a simple maximization context, becomes more problematic even in purely logical terms in game theory. This is so because a truly sophisticated player will not assume that his opponents will continue to stick to their old strategies if he varies his (either hypothetically or actually), i.e., he must take their reactions into account in some way. This sort of reasoning may lead one to superimpose plausible reaction functions onto the agents' strategy spaces, to look at hierarchies of leadership (Stackelberg solutions), or even consistent conjectural variations equilibria, in which the reaction functions themselves must satisfy certain rationality criteria. The latter leads to a sequence of CCV equilibria of increasing order (cf. Basar 1985) reminiscent of the Sherlock Holmes-Dr. Moriarty type cited by Morgenstern 1935.

The evolution of the notion of equilibrium in game theory underscores a fundamental problem confronting the rationality/equilibrium paradigm in orthodox economic theory. The fact that one may mathematically define a certain (or as in game theory a menu of such) consistency condition(s) in a complex interdependent behavioral system is still a long way from offering a compelling explanation for why the system should be in that state at any given time. There is something obviously artificial about selecting a solution concept and focussing attention on comparative static exercises with its help, merely because it can be defined and satisfies certain "attractive" (to the theorist) consistency relations. Reality may still blithely decide to remain inconsistent and go its own way, leaving the mess for historians to clean up. In the process, however, economists may be abdicating their responsibility to clarify the real by insisting all too exclusively on analyzing the "rational."

6. The 'As If' Classical Defense of Economic Equilibrium and the Competitive Process as an Explanatory Principle

Coexisting with the notion of a self-consistent equilibrium based on individual maximizing behavior is the concept of economic competition. These concepts are ordinarily considered by economists to be not merely mutually compatible but in fact mutually reinforcing. Competition supposedly punishes nonoptimal behavior, while competitive equilibrium ensures that no relative advantage can be attained by adopting a deviant strategy.

When it comes to reconciling this picture of the economic system with reality many of the most prominent neoclassical thinkers have been willing to admit that people rarely if ever perform the kind of explicit optimization calculations attributed to them in the theory. Does this then cast doubt on the validity of the whole approach? The answer given for example by Friedman 1953, Machlup 1946 and Machlup 1967, and Hahn 1984 (e.g. p. 58: "The reason why economists have for so long been interested in rational actions is because they claim that these have survival value.") is an interesting one, both for the bridge it (perhaps hypocritically) attempts to build to real-time models of learning and competitive selection, and for the conditions for equivalence to hold they somewhat cavalierly invoke. They argue that, by a process of economic natural selection, only those forms of routine behavior consistent with optimization will survive, even if they have become codified in rule-of-thumb operating procedures or implicit models of the world which do not obviously seem to correspond to the optimization description. Instead of going on to model explicitly this form of "disequilibrium" competition, however, most economists accepted this argument at face value as justifying a return to business as usual with a purified scientific conscience. Yet it is revealing that Friedman, Machlup and Hahn implicitly admit that the ultimate foundations of economic science must be located in an evolutionary theory of competition and not in a self-consistent equilibrium one deriving from an optimization theory of rational choice.

In many respects the issue revolves around the proper weight to be given to the Aristotelian causal categories brought up a number of times at the Abisko workshop in scientific explanations of economic pattern formation and change. The use of rationality in economics is strikingly reminiscent of Aristotle's conception of formal and final cause. Agents are driven by goal-seeking behavior, and a state of inconsistency between plans and outcomes is an "unnatural" one and thus abhorred by nature. In contrast, the evolutionary approach appeals to material and effective causes to explain causally the direction of change and the structure of interactions being worked out by the system at any point in time. Whether this dynamically converges to the finalist view of the economy is an open question which can only be addressed by formulating explicit models of evolutionary competition and realistic behavioral patterns which do not ex ante posit the limits to which they are striving. This is now fully recognized by the more sophisticated equilibrium theorists. (Viz., Smale 1980, p. 289: "... general equilibrium theory has not successfully confronted the question

'How is equilibrium reached?' Dynamic considerations would seem to be necessary to resolve this problem." See also Hahn 1984.)

But if the question of how equilibrium is attained is still open, the very notion of economic equilibrium as a privileged state in the dynamic is called into question. In some respects this finalist vs. effective causality debate resembles the alternative descriptions in physical optics between shortest path conditions based on variational methods and local partial differential equations governing wave propagation. Although the two descriptions can be shown to be mathematically equivalent (and thus more than just good approximations to each other), this is not at all obvious a priori.

Are we confronted with an analogous situation in economic theory? In a number of papers Sidney Winter has examined this question (Winter 1971, Winter 1975, Nelson and Winter 1982, Winter 1986). The problem initially is to formulate a dynamic model based on adaptive decision processes which can serve as a benchmark for comparison with the optimization paradigm. Without drawing on additional insights into the economic process this must remain an impossible task. In the simplist contexts, however, one may state that a competitive model may converge to the optimization equilibrium. However, this may be a considerably more time-consuming process than neoclassical theory implicitly assumes and thus would be invalidated if the boundary conditions did not remain constant for the required length of time. This calls into question the usual comparative static approach to the analysis of change, which does not distinguish with sufficient precision between the rate of change of boundary conditions and the rate at which adjustment to a new equilibrium can take place.

In the last few years the underlying structure of competitive processes and evolutionary selection has been the subject of renewed mathematical interest within the framework of dynamical systems theory. We are now approaching a state in which one may rightly speak of a certain systematization and clarification of the issues involved spanning a range of applications from molecular chemistry to sociobiology to laser physics (see in particular Ebeling and Feistel 1982, Hofbauer and Sigmund 1984, and Sigmund 1986 for overviews of the mathematical issues).

I will attempt to argue in the following that these same structures, suitably modified and interpreted, can shed some new light on this ongoing debate about the fundamental principles of economic explanation. The next section will then go on to search for a new set of robust patterns which will emerge from this critique of orthodoxy.

The fundamental equations of evolutionary selection can be cast in the form

$$\dot{f}_i = A(E_i - \langle E \rangle) f_i, \quad i = 1, 2, \ldots n, \tag{1}$$

where f_i is the share in the total population (market) of the ith of n competing species (firms, technologies, countries, behavioral patterns, etc.), E_i is its relative competitiveness, A is a rate constant, and

$$\langle E \rangle = \sum_1^n f_i E_i. \tag{2}$$

The competitiveness variables may represent measures of relative fitness (in terms of net reproduction rates) of populations or price/quality differentials in economic markets. In the equation as originally introduced by R.A. Fisher in 1930 (Fisher 1930), the E's were constants characteristic of different genetic variants of a single species. The behavior of the system then reduces to a particularly simple form: the species with the highest E drives out all the rest, and $\langle E \rangle$ is a Lyapunov function for the dynamic process of global convergence.

If one regards competitiveness or the underlying fitness variables it reflects as the target function of an optimization exercise, then in this simple case evolutionary competition does represent a real-time parallel process to achieve it. Moreover, no assumptions about differentiability have to be made about the target function, and entrapment in local maxima can be avoided by stochastically injecting mutant species from time to time. A considerable literature has sprung up in the last few years dealing with applications of evolutionary algorithms to a variety of optimization problems (see e.g. Holland 1975, Schwefel 1981, Brady 1985 and several of the contributions in Farmer 1986). In contrast to the instant optimization postulated in neoclassical theory, this form of evolutionary optimization takes time as a real-life process driven by behavioral diversity and trial and error. But, as every operations research practitioner knows, the actual numerical solution of optimization problems of any sophistication with available methods is also time-consuming. And in direct comparison, evolutionary algorithms have shown themselves to be highly competitive, even more so in parallel implementations.

In Silverberg 1988 I critically review the evolutionary selection models which have been concretely put forward in economics in the last few years. The thrust of these models is that, provided sufficient variety of behavior is present, that behavior will be selected for and come to dominate which corresponds to a cost-minimizing (profit maximizing) solution for the given boundary conditions. At issue is less the mathematical formulation, which generally reduces to a variant of Fisher's Fundamental Theorem of Natural Selection, than the economic assumptions entering into the models about agents' behavior and in teraction and the representation of technology and market demand. It does not have to be reiterated here that reasoning by mere analogy has its dangers. Even if we have reason to believe that an underlying mathematical structure is applicable as a systems tools across disciplinary boundaries, it is still necessary to fill the black boxes with a sufficient amount of carefully formulated applications-specific content.

Be that as it may, there are at least two reasons why this harmonious picture of evolution cum optimization must be relativized. First, evolution is intrinsically a stochastic process. Since superior innovations/mutations enter the system in very small numbers, there is always a finite probability for them to die out at this stage. Jimenez Montaño and Ebeling 1980 demonstrate on the basis of a master equation formulation of the evolutionary process that the probability of extinction and the relative superiority of a mutant are inversely related: an innovation or deviant strategy must be better by a certain not insignificant amount to stand a chance of proving its worth. We can go a step further in this stochasticization of the universe and allow the rate of mutability or error-making itself to be subject to selection.

In a simulation study Allen and McGlade 1986a demonstrate that a population with an initial disadvantage and a small but finite rate of reproduction error will eventually displace a superior population which reproduces flawlessly. This extends our notion of optimality to include, alongside superior performance, optimum error-making and exploration as a possibly competing dimension of fitness. One implication of these explicitly stochastic approaches is that system diversity is an unavoidable and probably desirable feature of evolutionary systems. This stands in contrast to the apparent uniqueness of optimization and deterministic equilibrium solutions.

But even deterministic systems are now known to display much richer behavior in the nonlinear domain. This will be the case if the competitiveness variables E_i in Eq. (1) are no longer constants but more complex expressions, such as functions of the population shares f_i ("frequency-dependent selection"). Evolutionary games in which interactions between members of a population occur at random are generally of this type: expected payoff to strategy i is the sum of the payoffs resulting from the encounter of strategy i with strategy j (to the first order a constant) times the frequency of j in the population (see Maynard Smith 1982, Hofbauer and Sigmund 1984). Other examples are cyclic catalytic chains ("hypercycles") such as are thought to occur in the early stages of molecular evolution, representing the mutual reinforcement of information carriers (genes) and functions (enzymes) (Eigen and Schuster 1979). In general frequency dependence results in multiequilibria: depending on initial conditions (the basin of attraction the systems finds itself in), convergence will be to one of a number of points. Cyclic and possibly chaotic behavior are also possible under certain circumstances. Ebeling and Feistel 1982 have termed this behavior hyperselection, in contrast to simple selection with a unique asymptotic state described above.

The possibility of hyperselection calls the traditional idea of evolution as an adaptive and ultimately optimizing process into question. The simplest version of an optimizing system is an adaptive landscape with gradient dynamics, i.e., steepest ascent, as in classical mechanics. Evolutionary systems will only possess this property in the most trivial cases. Sigmund 1985 shows, however, that by replacing the Euclidean metric by a particular Riemannian metric, the so-called Shahshahani metric, a gradient dynamics can be recovered for a larger class of systems. The Shahshahani metric give preferential weighting to states of the systems where some of the population frequencies are small. Even if a Shahshahani gradient description is possible, however, the population's average fitness need not always increase over time (in contrast to simple selection) if the gradient is not a homogeneous function of the frequencies.

This latter observation raises the question of extremal principles and target functions in the description of evolution, both at the system level (organic evolution as maximizing energy throughput, as Lotka proposed) and with respect to individual strategies. Ebeling and Feistel 1982 pp. 201–204 provide an exhaustive classification which differentiates in particular between static and dynamic extremal principles. A static extremal principle is defined on the fixed points and assumes a maximum value at one of them. A dynamic extremal principle is defined over the entire space and defines a Lyapunov function for each of the fixed points. Ebeling and Feistel call an extremal principle complete if it induces both a dynamic and a static extremal principle on the system. The converse of this classification is that there are systems which do not satisfy some or all of these criteria. In particular this implies that there are evolutionary systems in which, under the action of selection of "superior" types, average performance may actually decline for a while. It also implies that, depending on initially conditions, an absolutely superior strategy may still be driven out of existence and the system will lock in robustly to a different constellation.

The deterministic approach to hyperselection places the burden of choice of ultimate outcome on initial conditions. This of course is rather unnatural, since "initial conditions" are only a snapshot of the system at some arbitrary point in its history. From an evolutionary point of view, however, initial conditions can be reinterpreted as representing the early phases during the introduction of an innovation/mutation subject to large stochastic fluctuations. The choice of basin of attraction is thus shifted to the early pattern of historical events, a more or less random or idiosyncratic process which is gradually dominated by the deterministic constraints of the dynamics of large numbers. This kind of process has been termed path-dependent. A Strong Law for dependent-increment processes has been proven which generalizes the conventional Strong Law of Large Numbers for this class of stochastic processes (see Arthur, Ermoliev and Kaniovski 1984). Instead of assuming independence of random events in a sequence, the probabilities of each outcome are now assumed to be dependent on the frequencies realized up to that point of time (i.e., as a generalized Polya urn scheme). Any competitive process in which the probability of adoption of a strategy/technology increases with the number of previous adopters (e.g. due to network externalities, compatibility, imitation), or in which "fitness" or efficiency increases with use ("learning-by-doing or using" or increasing returns to scale) falls under this category and is thus analogous to frequency-dependent selection in the deterministic description. Arthur 1988 provides an introduction to the mathematical ideas and the main applications in the economics of competing technologies.

In these cases in which an unequivocal static or dynamic extremal principle cannot be defined, the notion of agents as utility maximizers is called into question. For the payoffs to evolutionary games are in terms of survival, growth or extinction to which, in these cases, no simple numerical measure can be assigned. Here utility maximization and competition may part company as explanatory principles in economic science. Profit under capitalist competition, as Keynes conjectured, may sometimes be more readily made not by correctly assessing underlying facts but by correctly anticipating public opinion. And the unresolved question of whether to maximize profits over the short or the long term, or with respect to some other target variable (market share, rate of growth), is another indication that evolutionary strategies may not be reducible to maximization problems.

In contrast to the rationality/equilibrium approach, all of these evolutionary approaches are both explicitly dynamic and do not assume away the diversity of interacting populations. This must be contrasted with the frequent recourse to "representative agents" and the sidestepping of the "aggregation problem" in most neoclassical models, as well as the invocation of "rationality" to compress disequilibria over time into a static framework. I would like to bring up briefly a third issue which is seldom mentioned in this connection and which I shall call the question of structural stability in the widest sense (to borrow a term first introduced apparently by Andronov and later developed by Thom in the study of dynamical systems). By this I mean the robustness of the conclusions of a model with respect to small variations in the parameters, the form of the equations, or even some of the behavioral assumption on which it is based. This is a well developed concept in the mathematics of dynamical systems. In economics the concept—usually in other guises—has only been applied in a few isolated instances.

Zeeman 1979 and 1981 has applied the concept of structural stability to evolutionary games by adding the replicator dynamics described above to the static formulation based on Evolutionary Stable Strategies (ESS) first proposed by Maynard Smith and Price. The first result is that ESSs, while point attractors, are not the only possible point attractors of the flow. Moreover, the global properties of ESS and non-ESS attractors differ. The second result concerns the properties of stable games, i.e., games whose topological properties do not change when the entries of the payoff matrix are slightly perturbed. Focussing on stable games restricts the kinds of bifurcations which can occur in parameterized games. In Zeeman 1981 the structural stability approach is used to show that a widely studied model of animal conflict is not stable; when the game is stabilized by slightly altering the payoffs, the previously "best," but nonattracting "retaliator" strategy does indeed become an ESS. However, a new attracting strategy also emerges in the process: a mixture of "hawks" and "bullies," which can be interpreted as a mixed strategy. Thus we are confronted once again with an example of the ubiquitous phenomenon of hyperselection. Moreover, the application of the concept of structural stability in this case yields unintended dividends in the form of Zeeman's hypothesis that this second ESS could explain the origin and stability of pecking orders in animal societies.

There are a number of scattered attempts, particularly in the game-theoretic literature, to address similar questions. Because they are not based on dynamic models they unfortunately cannot draw on the available deep topological results on structural stability in the original sense of the concept. I can only mention in passing a few approaches here. One is the competence-decision gap critique of classical choice theory by Heiner 1988, in which it is recognized that, even if agents are able in theory to identify the optimal action, there is always a certain probability of making a false move anyway (pushing the wrong button). This forces agents to rely on more robust rule of thumbs. Apparently related are such notions as the "trembling hand" and game models with a small percentage of finite automata players operating on mechanistic rules. A paradoxical result of these analyses is that the collectively suboptimal results of the fully rational equilibria can be shifted to the Pareto optimal frontier with the introduction of these kinds of irrationality in certain non-zero-sum games. There may be other reasons for microdiversity leading to major system- level changes. Binder 1986 for example considers indivisibilities of consumer purchasing decisions with varying tastes, maintaining the assumption of maximization of a utility function. A new classical econometrician coming to the problem based on the representative consumer would estimate a completely erroneous demand curve. Lippi 1988 also demonstrates that aggregation in autoregression models can lead to a completely misleading dynamic specification.

There are other dimensions in which the structural stability of classical results can be tested. One relates to the informational requirements of decision theory based analyses. Thus agents may have unequal access to relevant information (such as buyers and sellers), which has lead to the study of asymmetric information games. If we push the rationality notion of behavior under uncertainty to the limit and examine Bayesian games (in which players form and revise subjective probabilities about their opponents strategies and the state of the world), then as Aumann 1987 points out, departure from the implicitly made Common Priors Assumption calls convergence to an equilibrium fundamentally into question. Another dimension is the costliness of not being optimal. Akerlof and Yellen 1985 point out that small deviations from the theoretically optimal behavior may only have consequences to the individual of second order in the deviation. If some fraction of the agents are near rational in this sense, the change in the system response may be first order in the deviation, i.e., aggregate effects behave significantly more sensitively to deviations from equilibrium rationality than the costs to the individual.

At this point the unambiguousness of rational equilibrium behavior begins to break down, for if a significant number of players can be

expected to obey other rules, it is no longer rational for even the agent
most committed to optimization to adhere to the old strategy: he would
now have to take the "irrationality" of some of his opponents into
account. The notion of rationality in economics is both a worst case
and a utopian scenario. Worst case because it assumes others will
always faultlessly play for their own maximum egoistic advantage, and
utopian because it predicates information and computation ability not
accessible to mortals or machines (as the irrelevance of the minimax
solution to such a well-defined game as chess amply demonstrates).
If the equilibrium solution concept begins to crumble with the first
ϵ-order departures from its assumptions, however, it rapidly ceases to
provide any guide to the actual patterns of economic behavior.

7. Routine and Non-Routine Behavior, the Process of Collective Search and Schumpeterian Competition

The 'as if' defense of the neoclassical approach often makes an appeal
to routine or steady-state behavior. This is not an implausible assumption in the absence of fundamental uncertainty and major innovations
of an unpredictable kind. In Silverberg 1987 I attempt to address this
question in an evolutionary setting, examining the emergence of stable
rules of thumb for investment behavior when the rate of technological
change is more or less predictable. Evolutionary optimal strategies are
shown to depend on the nature of financing open to firms as well as
their price margins. In a case such as this it is highly plausible that
the results of evolutionary selection coupled with learning and imita-
tive behavior are sufficient to produce a coherent behavioral pattern
if boundary conditions remain more or less constant for rather long
periods of time. Axelrod 1984 also presents a fascinating evolutionary
approach to the emergence of stable cooperative behavior in human
and animal societies.

   The relationship between genetical learning (biological evolution)
and individual, neural learning as ordinarily understood is discussed in
Maynard Smith 1986 and Harley 1981 using the concepts evolutionarily
stable learning rule and rules for learning evolutionarily stable strategies. Harley derives properties for the evolutionarily stable learning
rule which makes it strikingly similar to notions of adaptive learning
or expectations. In equilibrium the rule must satisfy

$$\text{prob of choosing } A > \frac{\text{total payoff so far for playing } A}{\text{total payoff so far}}.$$

It will never abandon a strategy completely (and thus remain resilient) and will weigh recent experience more heavily than older experience. Behavior should change rapidly if experience is below expectation and more slowly if it is above it. However, a learning rule will only be sensible if the cumulative payoffs after equilibrium are attained are significantly higher than the payoffs during the learning phase. As Maynard Smith says, "if you are only going to play a game a few times, you had better be born knowing how to play it."

What, however, if we are outside this regime of routine behavior, where, in Schumpeter's words, "events have time to hammer logic into the minds of men?" In such cases it is more than doubtful that agents have anything but the vaguest notions about what their payoff matrix looks like. Their task then is both to explore this payoff space (which may be of extremely high dimension, very jagged, and historically unique) and stay in the evolutionary game. In terms of technical change this means not merely keeping up with developments originating from outside, but also choosing the right avenue for partly autonomous, partly collective development based on expectations of future potentials and the choices of other agents.

In Silverberg, Dosi and Orsenigo 1988 we argue that situations of this kind mandate behavioral diversity based on the technological "aggressiveness" or "conservatism" of the agents, attitudes whose a priori appropriateness cannot be dismissed out of hand. This is completely analogous to Schumpeter's innovating and imitating entrepreneurs. The presence of learning-by-using positive feedbacks and informational spillovers between firms makes the technological system highly path-dependent and collectivistic. Only if microdiversity is present will the collective exploration and exploitation of technological opportunities come about. Yet the existence of winners and losers is practically unavoidable. What keeps the system going is the uncertainty about which behavior will in fact yield a dividend. Figure 3 summarizes the results of a number of simulations for a particular distribution of firms' technological aggressiveness (how early they will adopt a new, immature but potentially superior technology) and for different values of the rates of learning-by-using and spillover. Noteworthy are the two bifurcation curves separating the region of failed diffusion, of relatively more successful imitation strategies and more successful early innovation. On the one hand it is the uncertainty concerning the exact values of the game parameters that leads agents to take diverse positions, and on the other it is the diversity of the positions taken which allows the diffusion process to get underway efficiently. Evidently the system

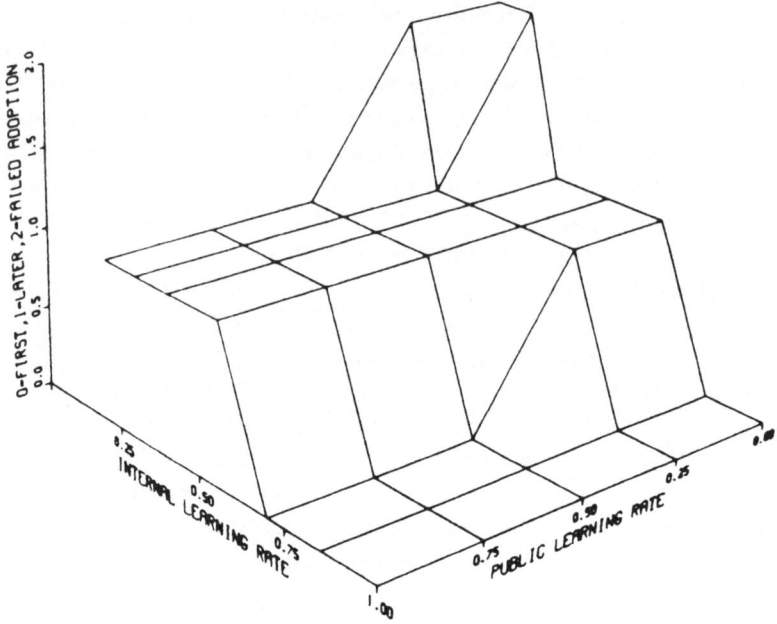

Figure 3. Net Winner of Diffusion Process as a Function of Rates of
Learning-By-Using and Public (Spillover) Learning

thrives on having both despite the unavoidability of "disequilibrium"
losses and gains in the process. The certainty of facing irreversible
decline if no anticipatory position is taken while others do is sufficient
to prod agents eventually into taking inherently risky action.

The notion of evolutionary disequilibrium as the driving force in
processes of development and exploration has also been taken up in a
number of other related contexts. Allen and McGlade 1986b for example analyze the dynamics of fisheries in terms of the interaction of
"stochasts," or high risk taking fishermen, and "cartesians," or imita-
tors who only sail to places of certain catch. Kwasnicka and Kwasnicki 1986 have also related evolutionary diversity to an alternation of
searching phases and gradual improvement phases resembling punctuated equilibrium. Evolutionary progression is predicated in an essential
way on a distribution of types, which, while lowering average fitness,
allows switch points to be identified and explored. Diversity increases
at such branching points, only to decline but not disappear during the
amelioration phases. This suggests that the specific patterns associated
with evolutionary processes are not self-consistent rational equilibrium

but rather steady-state or fluctuating population distributions emerging from the nonstationary motions of the individuals.

References

Akerlof, G. and J. L. Yellen, "Can Small Deviations From Rationality Make Significant Differences to Economic Equilibrium?," *Amer. Econ. Rev.,* 75 (1985), 708–720.

Allen, P. M. and J. M. McGlade, "Optimality, Adequacy and the Evolution of Complexity," paper presented to the MIDIT workshop *Structure, Coherence and Chaos in Dynamical Systems,* Lyngby, Denmark, 1986.

Allen, P. M. and J. M. McGlade, "Dynamics of Discovery and Exploitation: The Case of the Scotian Shelf Groundfish Fisheries," *Can. J. Fisheries Aquatic Sci.,* 43 (1986), 1187–1200.

Arthur, W. B., "Competing Technologies: An Overview," in G. Dosi et.al., 1988.

Arthur, W. B., Yu. M. Ermoliev, and Yu. M. Kaniovski, "Strong Laws For a Class of Path-Dependent Urn Processes," in Arkin, Shiryayev, and Wets, eds., *Proc. Int'l. Conf. on Stochastic Optimization,* Springer, Heidelberg, 1984.

Aumann, R., "Correlated Equilibrium as an Expression of Bayesian Rationality," *Econometrica,* 55 (1987), 1–18.

Axelrod, R., *The Evolution of Cooperation,* Basic Books, New York, 1984.

Basar, T., "A Tutorial on Dynamic and Differential Games," in T. Basar, ed., *Dynamic Games and Applications in Economics,* Springer, Heidelberg, 1985.

Binder, A. S., "A Sceptical Note on the New Econometrics," in M. H. Peston and R. E. Quandt, eds., *Competition and Equilibrium,* Philip Allan, Oxford, 1986.

Brady, R. M., "Optimization Strategies Gleaned From Biological Evolution," *Nature,* 317 (1985), 804–806.

David, P., *Technical Choice, Innovation and Economic Growth,* Cambridge University Press, Cambridge, 1975.

Davies, S., *The Diffusion of Process Innovations,* Cambridge University Press, Cambridge, 1979.

Dosi, G., C. Freeman, R. Nelson, G. Silverberg, and L. Soete, eds., *Technical Change and Economic Theory,* Pinter, London, 1988.

Ebeling, W. and R. Feistel, *Physik der Selbstorganisation und Evolution*, Akademie-Verlag, Berlin (DDR), 1982.

Eigen, M. and P. Schuster, *The Hypercycle*, Springer, Heidelberg, 1979.

Farmer, D. et. al., eds., *Evolution, Games and Learning. Models of Adaptation in Machines and Nature*, Special Issue of *Physica*, 22D (1986), Nos. 1–3.

Fisher, R.A., *The Genetical Theory of Natural Selection*, Clarendon Press, Oxford, 1930.

Fisher, J. C. and R. H. Pry, "A Simple Substitution Model of Technological Change," *Tech. Forecasting Soc. Change*, 3 (1971), 75–88.

Friedman, M., "The Methodology of Positive Economics," in *Essays in Positive Economics*, University of Chicago Press, Chicago, 1953.

Griliches, Z., "Hybrid Corn: An Exploration of the Economics of Technological Change," *Econometrica*, 25 (1957), 501–522.

Hahn, F., *Equilibrium and Macroeconomics*, Basil Blackwell, Oxford, 1984.

Harley, C. B., "Learning the Evolutionarily Stable Strategy," *J. Theo. Biol.*, 89 (1981), 611–633.

Heiner, R., "Imperfect Decisions and Routinized Production: Implications for Evolutionary Modelling and Inertial Technical Change," in G. Dosi et. al., 1988.

Hofbauer, J. and K. Sigmund, *Evolutionstheorie und dynamische Systeme*, Paul Parey, Berlin and Hamburg, 1984.

Holland, J. H., *Adaptation in Natural and Artificial Systems*, University of Michigan Press, Ann Arbor, MI, 1975.

Jimenez Montaño, M. A. and W. Ebeling, "A Stochastic Evolutionary Model of Technological Change," *Collective Phenomena*, 3 (1980), 107–114.

Keynes, J. M., *The General Theory of Employment, Interest and Money*, Macmillan, London, 1936.

Kwasnicka, H. and W. Kwasnicki, "Diversity and Development: Tempo and Mode of Evolutionary Process," *Tech. Forecasting Soc. Change*, 30 (1986), 223–243.

Lippi, M., "On the Dynamics of Aggregate Macroequations: From Simple Microbehaviour to Complex Macrorelationships," in G. Dosi et. al., 1988.

Machlup, F., "Marginal Analysis and Empirical Research," *Amer. Econ. Rev.*, 36 (1946), 519–554.

Machlup, F., "Theories of the Firm: Marginalist, Behavioral, Managerial," *Amer. Econ. Rev.*, 57 (1967), 1–33.

Mansfield, E., *Industrial Research and Technological Innovation*, Norton, New York, 1968.

Marchetti, C. and N. Nakicenovic, "The Dynamics of Energy Systems and the Logistic Substitution Model," IIASA Research Report RR–79–13, Laxenburg, Austria, 1979.

Maynard Smith, J., *Evolution and the Theory of Games*, Cambridge University Press, Cambridge, 1982.

Maynard Smith, J.,"Evolutionary Game Theory," in D. Farmer et. al., 1986.

Morgenstern, O., "Vollkommene Voraussicht und wirtschaftliches Gleichgewicht," *Zeitschrift für Nationalökonomie*, 6 (1935), 337–357.

Nelson, R. R. and S. G. Winter, *An Evolutionary Theory of Economic Change*, Harvard University Press, Cambridge, MA, 1982.

von Neumann, J. and O. Morgenstern, *Theory of Games and Economic Behavior,*Princeton University Press, Princeton, 1944.

Schumpeter, J., *Business Cycles. A Theoretical, Historical and Statistical Analysis of the Capitalist Process*, McGraw-Hill, New York, 1939.

Schwefel, H. P., *Numerical Optimization of Computer Models*, Wiley, Chichester, 1981.

Sigmund, K., "The Maximum Principle for Replicator Equations," in W. Ebeling and M. Peschel. eds., *Lotka-Volterra Approach to Cooperation and Competition in Dynamic Systems*, Akademie-Verlag, Berlin (DDR), 1985.

Sigmund, K., "A Survey of Replicator Equations," in J. L. Casti and A. Karlqvist, eds., *Complexity, Language, and Life: Mathematical Approaches*, Springer, Heidelberg, 1986.

Silverberg, G., "Technical Progress, Capital Accumulation, and Effective Demand: A Self-Organization Model," in D. Batten, J. Casti and B. Johansson, eds., *Economic Evolution and Structural Adjustment*, Springer, Heidelberg, 1987.

Silverberg, G., "Modelling Economic Dynamics and Technical Change: Mathematical Approaches to Self-Organization and Evolution," in G. Dosi et.al., 1988.

Silverberg, G., G. Dosi and L. Orsenigo, "Innovation, Diversity and Diffusion: A Self-Organization Model," *Economic J.*, to appear 1988.

Smale, S., *The Mathematics of Time. Essays on Dynamical Systems, Economic Processes, and Related Topics,* Springer, New York, 1980.

Tversky, A. and D. Kahneman, "Rational Choice and the Framing of Decisions," in R. M. Hogarth and M. W. Reder. eds., The Behavioral Foundations of Economic Theory, *J. Business,* 59, No. 4, Part 2 (1986), 251–278.

Winter, S. G., "Satisficing, Selection, and the Innovating Remant," *Quart. J. Econ.,* 85 (1971), 237–261.

Winter, S. G., "Optimization and Evolution in the Theory of the Firm," in R. H. Day and T. Groves, eds., *Adaptive Economic Models,* Academic Press, New York, 1975.

Winter, S. G., "Comments on Arrow and on Lucas," in R.M. Hogarth and M.W. Reder (eds), The Behavioural Foundations of Economic Theory, *J. Business,* 59 (1986), 427–434.

Zeeman, E.C., "Population Dynamics from Game Theory," in Z. Neticki, ed., *Global Theory of Dynamical Systems,* Springer, New York, 1979.

Zeeman, E.C., "Dynamics of the Evolution of Animal Conflicts," *J. Theor. Biology,* 89 (1981), 249–270.

Growth Cycles in Economics

ANDRÁS BRODY

Abstract

Three views compete concerning the behavior of a modern economic system:

1) According to Adam Smith (and the Nobel Laureates, most emphatically Debreu), the market moves to equilibrium.

2) According to Karl Marx (later Kalecki and Goodwin), the movement is cyclic with clearly observed periodicities.

3) Yet mathematicians, starting with Slutzky, deny both claims and, most shockingly with Mandelbrot, assert its erratic, chaotic, even monstrous behavior.

These inimical, even antagonistic views shall be reconciled as interwoven regimes of a simple model of economic growth. They may be various facets of the same process. Like knowledge itself, the human economy seems to grow in fits and jumps, in a haphazard, fluctuating but relentless manner.

1. Introduction

The aim of this study is to exploit the possibilties of large-scale input/output models for simulating the short and medium-term behavior of an economy, where production decisions rest on price information and price formation is governed by supply and demand.

The model draws on theories and ideas developed by Walras, Kalecki, von Neumann, Leontief, Sraffa, and Goodwin, and can also be considered as a purely theoretical investigation of a system of regulation: quantities acting on prices and prices acting on quantities.

The plan for the essay is simple: after developing the equations expressing price-quantity interactions, the resulting forms of economic motion are analyzed.

The notation will be the standard notation of input-output analysis: A = the flow coefficient matrix, B = the stock coefficient matrix, p = prices, and x = quantities.

2. Equations of Motion

This chapter develops a system of differential equations that may govern, at least in first approximation, the process of production in a detailed but closed economic system.

The setup of such equations may start from three different scientific principles. They are not only different but are usually considered contradictory. These three principles are as follows:

(1) the **optimizing** or normative approach, teleological in its essence, explaining processes backwards from their purpose or objective, i.e. their *causa finalis*;

(2) The **descriptive** or behaviouristic approach, accepting situations and features as they appear and deducing any rules only as they reveal themselves in actual situations;

(3) The **causal** or deterministic approach, inquiring into the motivation or the *causa efficiens,* and deriving the motion from seemingly *a priori* reasons or "forces."

Of these possible principles, the latter constitutes the classical way of reasoning, while the first is the general method of marginalism. Finally, the second represents the current positivistic, agnostic mode of handling scientific problems.

By deriving the model according to (1) and reinterpreting it independently in the light of (2) and (3), it will be shown here that in our context the three principles can be reconciled. The merit of the case will be the same whether the "entrepreneur" behaves as he does simply because he follows habitual behavior (as approach (2) asserts), or because he is maximizing his profits (as school (1) would put it), or owing to the "might of capital" represented by him as dumb tool of an unremitting process, plodding ahead. If the equations are justifiable according to all these approaches, we may hope to grasp a small slice of reality. The four Aristotelian principles of causation are evidently motivating the above approaches, with two of them—the "material" and "formal"—still mixed up in (2). To fulfill its descriptive, predictive, normative, and explanatory tasks, a model has to include or has to be built upon those Aristotelian principles. In modern terms they will figure as matter and energy, information and entropy—a closely connected set of basic axioms.

• Normative Deduction •

The usual procedure is the following: one announces the potential or objective function of the system, and by seeking its extremal value with the aid of differential calculus, one derives those equations that prescribe the optimal paths. Along these paths the system will continuously maximize its objective function.

In our case we may take as the objective the total, accumulated (not only transitory) gain produced by the whole system. We know that the transitory gain will depend at any given moment on the prices and quantities valid at that particular time, and we will look therefore for those rules that prices and quantities have to obey to insure a stable, global optimum. These "optimizing" prescriptions yield those "laws of motion" that every "entrepreneur," or simply every sector of production, has to comply with in order to achieve a long-run optimum and stability. This is indeed a hidden feasibility criterion for the system considered.

There are three elements of gain in any given situation. The expression of any of them has to admit variable, though strictly positive, prices and quantities. The first element is the profit realized on selling total production, or $p(1-A)x$, the difference of price and the flow costs of production. The second element is less obvious, and is usually called a windfall gain. It consists of the profit accruing from price changes in stocks and inventories. This gain can be expressed as $\dot{p}Bx$. It may sometimes be negative. The third element is a clear debit item: the outlays or costs of growth itself, the price of new stocks and inventories that have to be invested to increase the scale of production. This can be expressed as $pB\dot{x}$.

All three items have the dimension of flows and can thus be added in an objective function without introducing further dimensional constants. The objective function will therefore depend on p and x and will yield a scalar objective function that may be written in the form:

$$f(p,x) = \int_{-\infty}^{t} [p(1-A)x + \dot{p}Bx - pB\dot{x}]\, dt. \tag{1}$$

The integration shows that we are interested not in transitory but accumulated gains.

Before deriving a stationary curve for this function, a few observations are appropriate to characterize our particular choice of objective. The objective function maximizes the long-run difference of outlays and proceeds for the whole system. It does not start from the single, isolated entrepreneur who, by maximizing his individual profits, may be entirely indifferent toward the efficiency of the total system. The objective function thus presupposes an entrepreneur, who is usually assumed to behave in a more rational and common sense manner. His aims—if we discard private property relations and consider collectivist production—may just as well be accepted as the final aims of the whole community. In spite of this, it will yield a behavior that, at least in its essence and in its main tendencies, is strikingly similar to the behavior observed in an everyday private-property economy governed by the usual market relations.

The objective function is bilinear, maximizing a simple sum and not a rate, or a share, or a quotient. This leads to simpler mathematical forms that, moreover, can be easily interpreted. (Note: trials with other forms did not lead to appreciably better results.)

The objective function has only one not entirely "real" element, but it is indispensable. The component $\dot{p}Bx$ is theoretically "paper profit," a gain not backed by any actual physical surplus. Yet in everyday practice this paper profit cannot be distinguished from real profit, and a sizeable part of capital is always tied up in precisely these "speculative" domains. The price differential of inventories can usually be represented by credit operations backed with the "security" offered by the "high value" of inventories. In a market economy the intrinsic value of inventories is seldom important. The basic question is the actual price they fetch on the market. Without this item in the objective function, the resulting solution would still be cyclic. The only difference would be that the order of magnitude of the fluctuation, its amplitude, would be much smaller.

Finally, we are integrating over transitory gains. We know that "instantaneous" maximization of profits without employment of this long-run approach causes severe instability. To explain this in the simplest fashion possible, if we want to maximize surplus at any single point of time, we do not invest at that moment at all because investment is an outlay that becomes a source of gain only at a later period, not yet envisaged.

Applying elementary calculus to Eq. (1)—that is, forming its derivative according to time and setting it equal to zero—yields

$$px + \dot{p}Bx = pAx + pB\dot{x}. \qquad (2)$$

This relation can be interpreted as follows: the total production plus the price increase of existing wealth must cover the flow and stock inputs of the same production.

In these stationary circumstances there is no gain (there do exist regular profits—this is not the usual "zero profit" assumption). But outlays and receipts are equal. Or, if we eventually include commodity money in the system, the possible difference between outlays and receipts is balanced by accumulation or decumulation of money stocks held in the respective sectors.

If there are no price changes, then Say's law is valid in a quite elementary and general way. But let us examine what kind of changes of prices and quantities are permitted by Eq. (2). For an optional p_i or x_k, we may write

$$-\dot{p}_i/p_i = (x_i - \sum a_{ij}x_j - \sum b_{ij}\dot{x}_j)/\sum b_{ij}x_j$$
$$\dot{x}_k/x_k = (p_k - \sum a_{jk}p_j + \sum b_{jk}\dot{p}_j)/\sum b_{jk}p_j. \tag{3}$$

Since we have to consider equations of this type throughout the remainder of this chapter, let's introduce a notational shorthand.

First of all, the Leontief-matrix $1 - A$ will be designated as C. The second convention is somewhat unusual. As we see, we have to consider the element-wise quotient of two vectors. A similar operation is employed when defining the matrix B as $\{b_{ik}\} = \{a_{ik}t_{ik}\}$, the element-by-element multiplication of two matrices. We have to find a convenient notation for these operations, which are not commonly found in matrix calculus. To this end, we define

$$a/b = c \quad \text{meaning} \quad c_i = a_i/b_i,$$
$$a \cdot b = c \quad \text{meaning} \quad c_i = a_ib_i, \qquad i = 1,\ldots,n.$$

It must be stressed that these operations are *not associative* as with usual matrix multiplication or inversion. Therefore $(Ax) \cdot y \neq A(x \cdot y)$, etc. But granted a certain care in manipulation, the notation is clear and self-explanatory. Eq. (3) can now be written as

$$\dot{p}/p = (-Cx + B\dot{x})/Bx,$$
$$\dot{x}/x = (C^T p + B^T \dot{p})/B^T p. \tag{4}$$

This is a non-linear system in $2n$ variables. On the left-hand side we find the logarithmic derivatives of the variables. On the right-hand side there are no "poles," i.e., no quotient grows infinitely large since as long as p and x remain nonnegative, the irreducibility of B guarantees strictly positive denominators. The logarithmic derivatives and the right-hand sides both suggest "relations" or "proportions" and express economic interdependencies that are indifferent to the choice of the units of measurement. Indeed, as inspection shows, the system is homogeneous of degree 0 and, as such, "measure invariant." By changing units of measurement separately for every x (and so also for p), the systems response does not change. Its economic content, i.e., the interpretation of the equation, will be taken up more thoroughly in the next section.

• Descriptive Explanation •

Let us put aside these equations of motion for awhile and consider the description of the process of regulation. We use only the basic input-output variables and build relations based upon clues offered by the language of everyday business. But instead of one commodity, let us apply the reasoning to n commodities, and instead of an undefined or implicitly static context, let us face the process of regulation in a growing economy.

Our "tool kit," well known in Input-Output Theory, consists of the following variables:

x	level of production
Ax	flow inputs
Bx	stock inputs
$B\dot{x}$	change of stocks
$A^T p$	costs
$B^T p$	capital/output ratios
$B^T \dot{p}$	change of same

Common sense gives general guidance as to how to piece together these variables and how to combine them in order to reflect the two kinds of decisions that "entrepreneurs" take: about prices and about production.

The first kind of decisions, the price decisions, are usually treated as if not the "entrepreneur," but some impersonal "market" takes them. But a mathematical model only states the impulses that lead to a change in any of the variables. It is therefore irrelevant to spell out "who" (the entrepreneur, the planner, or the "invisible hand") operates the variables.

"Excess demand leads to a rise in prices," argued Adam Smith. Price rise is $\dot{p}$, but what is excess demand? Demand in a growing economy has two components: the demand generated by current production and an additional quantity, usually not constant, needed for growth itself. On the supply side, we have the quantity produced x, and on the demand side we have to consider not only the flow demand Ax needed to maintain this level of production, but also the additional demand $B\dot{x}$ resulting from the growth decisions. Now let us heed the admonitions so often stated by our primary school teacher: do not add apples to oranges. The question is how to make the two sides commensurable. Intuition (not forbidden even in a behaviouristic description) tells us that it is relative changes and relative quantities we are after. An unsold bag of maize will presumably have a lesser effect on the Chicago grain market than in a remote African village. So instead of the simple time derivative, we take the logarithmic derivative of prices, $\dot{p}/p$, which has the beneficial effect of rendering the expression independent of the particular currency, dollars, pounds, marks or yen.

On the other side of the equation we also have to look for a relation. One could relate excess demand to the total quantity produced, i.e., to x. But on second thought, this would be a dimensional mistake since the latter is also a flow. To avoid dimensional trouble, we have to divide by stocks since the derivative on the price side is of the dimensionality time^{-1}. The only stock quantity we have in our modest tool kit is Bx. We thus arrive at a market equation having the form:

$$\dot{p}/p = (Ax + B\dot{x} - x)/Bx.$$

This means that the relative change in prices is governed by relative excess demand. The equation, incidentally, is identical to the first part of the dual system as set out in Eq. (4).

Can one defend the particular choice of the divisor Bx? In qualitative terms, certainly: all goods produced and not yet consumed are for sale; therefore, "on the market." Quantitatively it is more difficult to argue. This would be the proper place to introduce some "behaviouristic parameters" that connect price reactions to the state of supply and demand and correct it, differentially, for each separate commodity. By sleight-of-hand we discarded all of them, or rather replaced them with the reciprocal of the stock vector.

Nevertheless, in a descriptive context one can make the point that if no such parameters are introduced, then no further information is needed. And, indeed, the estimation of the parameters would tremendously increase the statistical effort needed to build up the model. Within the very limited assortment of observational material at our disposal, we obviously made a very economic choice.

The case, *prima facie,* is not against our choice. In business cycles, where excess demand (or supply) fluctuates with the fluctuating general rate of growth (that is, in parallel with $B\dot{x}$), this fluctuation, when divided by Bx, will produce fairly uniform relative amplitudes in price changes—and this coincides with actual observations.

Prices and their changes now govern the dual decisions—those about the scale of production. So we embark upon the production side bearing in mind that according to Adam Smith and David Ricardo, "high profits prompt the dealers to employ more labour and capital."

We have here a relation from the outset, high profits standing evidently for a high rate of profits, not simply for a large amount of profit realized on sales in itself, but in relation to the capital advanced. Within the confines of our model, where no flow of funds, i.e., no "transference of capital," is considered, it must be this rate of profit that governs the decisions to augment or decrease production. Hence it would be natural to equate the rate of profits with the percentage increase of production, again a logarithmic derivative, this time that of the levels of production $\dot{x}/x$.

This description of the actual production decisions tallies with rough statistical facts. Profits and growth rates of industries do show a strong correlation whether run as time-series or as cross-section analyses. This is almost a truism since once "transference of capital" is neglected (and this transference itself will be regulated also according to the particular rate of profits), the profits in a given branch, or on a given commodity, allow exactly the same percentage increase if spent on additional stocks. With constant technology, as we have it, we seemingly could have no better approximation to whatever happens.

But on second thought, a slight improvement seems to be possible. Our previous considerations missed an important aspect of the process by not incorporating the effect of price changes. We know that rising prices do affect the investment decisions favorably, and the decline of the general level of price does have an adverse effect. Yet the rate of profits itself, pCx/pBx, fails to reflect anything of this kind. It is insensitive to both the absolute level of prices and to the change of this level, because the same price vector figures twice and so cancels out the very effect we are looking for.

By taking into account the nominal appreciation (or depreciation) of stocks and inventories, i.e., $B^T\dot{p}$, we can take care of this problem. In essence these are windfall gains, unearned profits (though they will be mostly losses in slumps). They can be "realized" or converted into money only with the aid of the credit system. The increased value of inventories will be collateral for the banks, not to mention a basis for the kind of speculation that regularly places its money exactly where price rises may be expected. In correcting the entrepreneurs profits with this additional term, we probably underestimate its impact. Still, by use of this term we introduce into our descriptive model, without openly doing so, that function of money that is most intimately tied up with the cycle: earning without actually selling, or selling without being able to replace the items already sold.

On the other hand, as we will show in the next section, this plus (or minus) far from being virtual or just "paper profit," is exactly that amount of real increase (or decrease) in purchasing power that accrues to the producers as a mysterious addition in booms and as an equally sinister deficit in slumps.

Piecing together our previously considered quantities, we arrive at

$$\dot{x}/x = (p - A^T p + B^T \dot{p})/B^T p.$$

This equation means that the relative change in production is governed by profits, where profits are of two kinds: profits on sales and windfall gains (or losses). The equation, incidentally, is identical to the second part of the dual system set out in equation (4).

To sum up briefly, mirroring reality in a fairly simplified manner in the equations, we tied together two simple statements:

a) The relative price changes are equal to the relation of excess demand to the total stock of the respective products.

b) The relative increases in production are equal to the relative profits.

In both cases we dropped the possible additional proportionality factors, claiming those changes to be not simply *proportional* but *equal*, percent for percent. Equality leads to a simpler model. But there are also theoretical arguments for strict equality. We may stress that the possible inequality of $(1 - A)x$, the surplus, to $B\dot{x}$, the investment, leads to strongly felt tendencies in the system. These two quantities are seldom, if ever, equal. Still, in the long-run they must be equal. We cannot accumulate unproduced physical surplus, and surplus products, once produced, must be accumulated sooner or later.

The same applies to the dual side: monetary surplus or "savings" cannot constantly lag behind—or surpass—actual investment. If a sector operates with a constant loss, it consumes its capital sooner or later—and it seldom can operate under loss for long without curtailing production.

Seen from this aspect, which will be taken up again in the next section, the particular way our equations are connected is but a restatement of Say's law. It no longer claims that supply and demand must be equal in every instant of time, but maintains that they must remain equal or nearly equal when longer stretches of time are considered. The introduction of proportionality (instead of equality of relative proportions) would blur this important interdependence.

There is still an important omission in our equations when considered from this aspect. The equations do express the effect of, say, excess supply on prices. But they fail to give a precise account of the excess itself. They do not show where it can be found, whose inventories grow, and how the excess possibly accumulates (or decumulates when excess demand is the order of the day).

Some nebulous "floating stocks," seemingly not belonging to anybody (and not costing anything to maintain) are therefore implicit in this formulation. I have not yet found an explicit solution for this problem that would not unduly complicate the basic equations. This excess stock is of the form $\int(1-A)x\,dt - Bx$, with the emergence of the integral form signaling the underlying difficulties. For the time being, we'll just have to neglect the problem.

The existence of such a "hidden reservoir" offers a certain flexibility in the model. There is the possibility to accumulate and invest somewhat more (or less) in any given instant of time than the surplus product produced at the same instant. Investment $B\dot{x}$ may surpass (or lag behind) surplus $(1-A)x$, by drawing on (or discharging into) that reservoir without further complicating the issue. With this reservation in mind, we now turn to the third approach: causal induction.

• Causal Induction •

To derive equilibrium prices and quantities, economic theory usually assumes only that they must render the continuation of the process (replacement or growth) possible. But if the system is not in equilibrium, it must still be able to continue its operation. Let us look, therefore, into the exigencies of the process: how must prices and quantities be modified to render the process feasible? Let us then investigate what changes in prices and quantities are possible and necessary in any non-equilibrium situation.

Profits make expansion possible. But what makes it necessary? I believe the final answer cannot be given on this level, where we neglect the technological change that steadily reduces the costs of production. The gradual devaluation of all existing wealth makes pure hoarding, if not impossible, at least extremely wasteful.

But already in our limited context a similar pressure makes itself felt: if the surplus at hand is not accumulated, the ensuing lack of demand (the missing "growth demand" $B\dot{x}$) will trigger a price decrease; hence, a loss on all the pre-existent stocks and inventories is unavoidable. To keep the price level and thus the already accumulated wealth intact, the surplus has to be accumulated.

This inner tension manifests itself outwardly as the "competition of entrepreneurs." Yet it is not the competition that gives rise to investment. Rather it is this possibility and necessity of growth that brings about the particular form of competition we observe, where those entrepreneurs will be most rewarded who comply most adequately with the inner necessities of the process.

We have to pause here because our argument took the usual reaction of market prices for granted. An excess demand makes price increases possible. Yet what makes them necessary?

The right answer can again be found only if we do not regard price changes simply as rewards (or punishments) of competitors but look into their deeper functions. Prices change because a disequilibrium is present, and they have to change in such a way to make further production possible despite the discrepancy of supply and demand.

The fundamental role of price changes is to reallocate the means of production toward increased production of the deficient commodities. Without a reallocation, the missing amount could not be forthcoming, nor could superfluous production be curbed. The need for reallocation determines not only the required direction of the price changes (an increase for the deficient and a decrease for the overabundant commodities), but also their magnitude. The price changes have to be such that the actual process of reallocation is rendered feasible.

Let us look into the question in a quantitative manner. The value (price times quantity) of supply in any given amount is $p \times x$. Here we use our new notation, not to be confused with the usual scalar product, to emphasize a relation that has to hold component-wise for every $p_i x_i$, and not only in the aggregate. The value of supply may be greater or less than the demand: $p \times B\dot{x} + p \times Ax$, i.e., stock and flow demand taken together. If there is a discrepancy, the price change must be in such a direction and of such a size that computed over all the existing stock of commodities (Bx), it should even out the discrepancy. Therefore

$$\dot{p} \times Bx = p \times Ax + p \times B\dot{x} - p \times x.$$

The equation obtained is again the market equation of system (4), written and interpreted in a different manner. If we envisage the dual equation in the same spirit, we will be led to the insight that even the investment decisions can be interpreted in a down-to-earth way. What seemed to be a sophisticated optimization according to our first approach, and an assessment of the respective rates of profits according to the second, now turns out as a much simpler activity where the entrepreneur just accumulates whatever surplus he is left with.

At a given moment his outlays on current account will be $A^T p \times x$. Other entrepreneurs will purchase from him, as already detailed above,

$p \times Ax$ on flow and $p \times B\dot{x}$ on stock account. Now if the above market equation holds, then all these purchases can be expressed as $\dot{p} \times Bx + p \times x$ by re-arranging the above equation. Therefore, the difference of his outlays and sales proceeds will be equal to $(1 - A^T)p \times x + B^T\dot{p} \times x$. And if he now divides this surplus by the capital intensity $B^T p$, he will arrive at that increase in production $\dot{x}$ which he can afford. Piecing things together, we arrive at

$$\dot{x} = ((C^T p + B^T \dot{p}) \times x)/B^T p.$$

The notation is somewhat involved, but we can divide through by x, whence we arrive again at the equation of the original system (4)

$$\dot{x}/x = (C^T p + B^T \dot{p})/B^T p,$$

with the very simple meaning worked out in our last approach: the entrepreneur spends what he receives.

This final conclusion is in an interesting contrast with Robinson's usual contention (attributed to Kalecki: "workers spend what they get, and capitalists get what they spend."). The entrepreneur here is considered less free but, then, he is not blamed for causing the cycle by stubbornly refusing to spend enough even though he could afford to. "Plethora of capital," unspent savings, do not exist at the height of the boom, nor at the turning point. What the model explains is how the entrepreneur is left without the necessary surplus to continue the general overproduction.

3. Forms of Motion

We prove first that a so-called "equilibrium" solution does exist and is unique. Luckily the two great schools of economic theories—labour theory of value and marginalism—defined "equilibrium" in a way furnishing identical magnitudes; they differ only in terminology and in the reasoning leading to their definitions.

What Marx called "prices of production"—prices that yield a uniform profit rate on the capital invested in the respective branches of production—marginalists defined as "long-run normal prices"—also yielding the same "normal" rate of interest on capital advanced.

What Marx demonstrated as stable proportions of production in the respective branches, growing uniformly to solve his tables of "extended reproduction," turned out to be the "turnpike path" or the "von Neumann-ray" of neoclassical writers, along which the economy can move smoothly without creating superfluous products or encountering unwanted bottlenecks. The strict duality of the growth rate and the rate of interest (which labour theory envisaged rather as the average rate of profits) has been first rigorously proved by von Neumann.

Inserting these "equilibrium" solutions into our equations, we find that in their neighbourhood only cyclic paths exist—some hundreds of thousands of them—because each product has its particular cycle with a characteristic frequency, phase, and amplitude. Then we return again to the equilibrium solution and consider the particularly nasty possibility of slipping and tilting sequences and, finally, relying on computer simulations, discuss the "grainy" structures of economic reality.

• Equilibrium •

Prices $\bar{p}$ yielding a uniform rate of profits in every branch on each and every product, that is $C^T \bar{p} = \lambda B^T \bar{p}$, if placed into the second equation of our system give rise to a uniform rate of growth. Hence, $\dot{x} = \lambda x$.

Do such prices exist? I offer two proofs. The first is the generalized Farkas-inequality which claims that if for every x entailing $Cx > 0$, we have $Bx > 0$, then there must be a positive $\bar{p}$ for which $C^T \bar{p} = \lambda B^T \bar{p}$. Expressed in economic terms: if surplus can only be produced by investing capital, then the existence of a positive price system is insured.

The second proof is somewhat longer. If there exists a positive x for which $Cx > 0$, then of course $x > Ax$, and therefore the spectral radius of the matrix A is strictly less than unity for productive systems. Therefore the so-called Leontief inverse,

$$C^{-1} = (1 - A)^{-1} = \sum_{1=0}^{\infty} A^i,$$

always exists and is strictly positive.

The matrix B being indecomposable, we know that the product $C^{T-1} B^T$ is positive and so possesses a Frobenius eigenvalue and eigenvector: a unique positive eigenvalue of maximal modulus to which a unique positive eigenvector belongs—our price system $\bar{p}$. This is also the simplest way to compute it: an iteration started with any nonnegative vector must converge to the solution. The uniform rate of profits is the *reciprocal* of the Frobenius eigenvalue—thus of the smallest modulus among the reciprocal eigenvalues.

The same proof, carried out for the matrix $C^{-1}B$, insures the existence of a positive vector of production $\bar{x}$ which can be augmented by the common growth rate λ. It maintains market-equilibrium, with supply and demand clearing smoothly and continuously, since $\lambda B\bar{x} = C\bar{x}$, or $B\dot{x} = C\bar{x}$ with $\dot{x} = \lambda\bar{x}$. The surplus produced is equal to the capital invested.

If these proportions are placed into the first equation of our system, then the ensuing $\dot{p}/p$ will be zero, because the universal equilibrium will ensure unchanging prices. We found, therefore, a particular solution triple for our system $(\lambda, \bar{p}, \bar{x})$ that is unique and economically meaningful.

Two features have to be stressed. The vectors $\bar{p}$ and $\bar{x}$ are not dual vectors because they belong to different (but similar) matrices. The left eigenvector of the system $C^{T-1}B^T$ yields equilibrium capital proportions $B\bar{x}$. The left eigenvector of the system $C^{-1}B$ expresses the equilibrium capital intensities $B^T\bar{p}$. If we want to consider $\bar{p}$ and $\bar{x}$ in strict duality, we have to form the matrix-pencil $A + \lambda B$. Then they will be those left and right eigenvectors which belong to the *unit* eigenvalue of the latter matrix. But, as already noted, other numerically feasible but non-positive solutions will yield higher values for λ; thus the stability of the system is endangered whenever we strive for simple maximization.

•   Cycles   •

Taking $\bar{p}$ and $e^{\lambda t}\bar{x} = \bar{z}$ as our previously found solutions and inserting them into the equation system (4), we can transform it to the following hypermatrix form:

$$\begin{pmatrix} B\bar{z}/\bar{p} & -B \\ -B^T & B^T\bar{p}/\bar{z} \end{pmatrix} \begin{pmatrix} \dot{p} \\ \dot{z} \end{pmatrix} = \begin{pmatrix} O & -C + \lambda B \\ C^T - \lambda B^T & O \end{pmatrix} \begin{pmatrix} p \\ z \end{pmatrix}. \qquad (5)$$

Upon inspection we find on the left hand side a symmetric $S^T = S$ matrix, while on the right-hand side a skew-symmetric matrix $K^T = -K$. With the notation $r = (p, z)$, we can express these features simply as

$$S\dot{r} = Kr. \qquad (6)$$

Such systems have only zero and purely imaginary eigenvalues, so their behavior cannot be anything but cyclic. The proof is again simple. Assume an eigenvector of the form $(a + ib)$, with $i^2 = -1$, and an eigenvalue $(\alpha + i\beta)$, i.e.,

$$(\alpha + i\beta)S(a + ib) = K(a + ib). \qquad (7)$$

Premultiplying now by $(a - ib)^T$, we note the following consequences: with the symmetric matrix on the left side, the imaginary part cancels since

$$(a - ib)^T S(a + ib) = a^T Sa + b^T Sb.$$

With the skew-symmetric matrix on the right side, the real part cancels since

$$(a - ib)^T K(a + ib) = 2ia^T Kb.$$

Having a real quantity on the left side and an imaginary on the right, α must cancel in Eq. (7), i.e., $\alpha = 0$, and $\beta = 2a^T Kb/(a^T Sa + b^T Sb)$.

We can simplify matters by premultiplying Eq. (7) by $(a + ib)^T$, whence, because the right-hand side cancels, we arrive at

$$i\beta(a^T Sa - b^T Sb + 2ia^T Sb) = 0.$$

We conclude $a^T Sa = b^T Sb$ and $a^T Sb = 0$, because both the real and the imaginary part must equal zero. Therefore, finally

$$\beta = a^T Kb/a^T Sa = a^T Kb/b^T Sb, \tag{8}$$

will yield the respective cycles. None will be "synchronous." In other words, they will not occur with the same phase because a common phase would require $a = \rho b$, which immediately entails $\beta = 0$ by virtue of the skew symmetry of K. A synchronous cycle can belong only to the singular solution of Eq. (6), where both sides equal zero. We know this solution. It is the "equilibrium" of the former paragraph, which is unique. But is it truly unique?

• Nasty Disturbances •

The situation is already grim enough: we have a growth path, stable, but not assymptotically stable, with superimposed asynchronous cycles, several hundreds of thousands of them, as every branch, nay, every product, triggers its own particular periodicity, phases, and amplitudes. I tried to simulate such trajectories on a computer and when the number of randomly generated frequencies is above the usual "small sample size" of thirty, it is impossible to distinguish it from Gaussian noise. If this periodicity sits in the exponent, the noise will be lognormal, which again can be distinguished from a hyperbolic distribution only at the tails—which are cut off in a finite sample.

To add to our problems, a further curious feature emerges. Economists like to speak about matrices and vectors as simple scalars in the same way they like to "aggregate." And though aggregation is not always tidy, we can use it here to present this curious feature: a possibility for a slip in the equilibrium. Let us assume, therefore, that our economy turns out just one good and has just one price. The vectors x and p are therefore scalar quantities, also the matrices A and B. Let us even assign them approximate numerical magnitudes: 0.9 for A, hence 0.1 for C and 2 years for B. The equilibrium growth rate $C/B = 0.05$, or five percent/year.

Our system now is the following:

$$\begin{pmatrix} 2x/p & -2 \\ -2 & 2p/x \end{pmatrix} \begin{pmatrix} \dot{p} \\ \dot{x} \end{pmatrix} = \begin{pmatrix} & -0.1 \\ 0.1 & \end{pmatrix} \begin{pmatrix} p \\ x \end{pmatrix}. \tag{9}$$

With $\dot{p} = 0, \dot{x} = 0.05x$, both sides yield $\begin{pmatrix} -0.1x \\ 0.1p \end{pmatrix}$ and are therefore equal.

The matrix on the left side is of course singluar. Yet if there is a small disturbance, say ϵ, in the measurement of capital intensity, then the matrix can be inverted, and indeed

$$\begin{pmatrix} 2x/p & -2 \\ -2 & 2p/x + \epsilon \end{pmatrix}^{-1} = p/2\epsilon x \begin{pmatrix} 2p/x + \epsilon & 2 \\ 2 & 2x/p \end{pmatrix}.$$

Thus the system (9) can be transformed into

$$\begin{pmatrix} \dot{p} \\ \dot{x} \end{pmatrix} = p/2\epsilon x \begin{pmatrix} 0.2 & -0.2p/x - 0.1\epsilon \\ 0.2x/p & -0.2 \end{pmatrix} \begin{pmatrix} p \\ x \end{pmatrix}. \tag{10}$$

But this system yields a different solution: $\dot{x} = 0, \dot{p} = -0.1\epsilon x p/2\epsilon x = -0.05p$. The solution is independent of the magnitude of ϵ which cancels.

And, indeed, if production stagnates, then the unneeded surplus accumulating will depress the price level by just that amount which leads to the stagnation, because it seemingly wipes out the profit and the anticipation of profits after any new investment.

Returning now to the fully fledged economic system (4), we observe that though the solution triple $(\lambda, \bar{p}, \bar{x})$ is truly unique, behind every growing economy—p stable, x growing at the rate λ—there lurks an "anti-economy"—x stagnating, p decreasing at the rate λ—and this anti-economy can be triggered by the smallest disturbances to take over the "real" economy.

And disturbances will be numerous, not only because of the cycles already mentioned, but also because of the very nature of economic processes. I believe the problem here rests mainly in the following: *stocks*, amounts of commodities, can be observed and measured correctly, but *flows* are curious figments of economic science. They are artificial and, strictly speaking, unobservable phenomena not only because certain products—say an automobile or a blast-furnace—are indivisible, but because the production of even the most finely divisible products—say pig iron—is never continuous. The economic day and week and month and year is intermittent—generating intermittent streams of products. By averaging over monthly or quarterly or yearly production, we push this grainy structure of production into the background. To grasp economic reality, we shall need new mathematical tools that can express the discontinuous, discrete character of the basic production processes.

This character of interdependencies was seen also in the computations I made with the system (4). When the steps were large, of size 0.1, indicating monthly decision-making, the system behaved atrociously and even had a tendency to diverge. By reducing the step-size below 0.01, the fluctuations persisted but without the tendency to ruin the trajectory: 20–30 years of simulated forecasts could be easily made. My guess for a realistic step size is, however, smaller, about 0.003, indicating a daily decisionmaking routine. I believe that if the economic situation becomes tight, the decisionmakers' do become more alert and flexible. In real life, the "step size"—that is, the frequency of decisionmaking—depends on circumstances. If sailing is smooth, managers go on extended vacations; in crisis times, the staff is in on duty evenings and weekends.

References

Brody, A., *Proportions, Prices, and Planning*, North-Holland, Amsterdam, 1970.

Goodwin, R. M., "A Growth Cycle," in A. Feinstein, ed., *Socialism, Capitalism, and Economic Growth*, Cambridge University Press, Cambrdige, 1967.

Goodwin, R. M., "Swinging Along the Turnpike with von Neumann and Sraffa," *Cambridge J. Econ.*, 10 (1986), 203–210.

Kalecki, M., "A Macrodynamic Theory of Business Cycles," *Econometrica*, 3 (1935), 327–344.

Leontief, W., "Structural Matrices of National Economies," *Econometrica,* 17 (Supplement) (1949), 273–282.

Modeling Language Change:
Ontogenetic and Phylogenetic

DAVID W. LIGHTFOOT

Abstract

Children develop their linguistic capacity along lines which are prescribed genetically. Given rich genetic prescriptions, exposure to a particular linguistic community determines the developmental stages that children go through. This model of language acquisition casts new light on how one might explain and model how language systems change historically from generation to generation.

1. Introduction

In this paper I shall first discuss how children acquire their native language, arguing that the process is facilitated by information which is provided genetically. This information consists of principles and parameters, and children develop a mature linguistic capacity by setting the parameters when exposed to some linguistic environment. That will put us in a position to ask how a linguistic environment might change in such a way as to entail a new parameter setting and thus how languages change from one generation to the next. I shall argue that change proceeds in a gradual, piecemeal fashion, punctuated with occasional sudden, large-scale, "catastrophic" changes. This phenomenon is interesting for models of biological change because the human linguistic capacity manifests one component of our genetic make-up which is peculiarly open and plastic, consistent with the phenotypic state of a Chinese-speaker and that of an Italian-speaker. How classes of phenotypic states replace one another imposes special requirements on development models. First, we must establish that language development in a child is subject to genetic prescriptions.

2. Ontogeny

For several years generative grammarians have been developing a selective theory of language acquisition. We have sought to specify relevant information which must be available to children independently of any linguistic experience, in order for a particular child's eventual mature capacity to emerge on exposure to some typical triggering experience. Cutting some corners, we have assumed that that information is genetically encoded in some fashion and we have adopted the explanatory model of (1).

1a. trigger (genotype ⟶ phenotype)
1b. primary linguistic data (universal grammar ⟶ grammar)

The goal is to specify relevant aspects of a child's genotype such that a particular mature state will emerge when the child is exposed to a certain triggering experience, depending on whether the child is raised in, say, a Japanese or Navaho linguistic environment. (1b) reflects the usual terminology, where "universal grammar" (UG) contains those aspects of the genotype directly relevant for language growth, and a "grammar" is taken to be that part of a person's mental make-up which characterizes his or her mature linguistic capacity.

The theory is **selective** in the same sense that current theories of immunology and vision are selective and not instructive. Under an instructive theory, an outside signal imparts its character to the system that receives it, instructing a plastic nervous system; under a selective theory, a stimulus may change a system by identifying and amplifying some component of already available circuitry. Put differently, a selective theory holds that an organism experiences the surrounding environment (and selects relevant stimuli) according to criteria which are already present. Jerne (1967) depicts antibody formation as a selective process whereby the antigen selects and amplifies specific antibodies which already exist. Similarly Hubel and Wiesel showed that particular neurons were pre-set to react only to specific visual stimuli, for example to a horizontal line; there follows a radical increase in the number of horizontal line receptors and a horizontal line can be said to elicit and select specific responses within the organism. Changeux (1980, 1983) argues along similar lines for a theory of 'selective stabilization of synapses' whereby 'the genetic program directs the proper interaction between main categories of neurons ... However, during development within a given category, several contacts with the same specificity may form' and other elements, which are not selected, may atrophy (1980 p. 193). Thus to learn is to amplify certain connections and to eliminate other possibilities. Jerne argues that 'Looking back into the history of biology, it appears that wherever a phenomenon resembles learning, an instructive theory was first proposed to account for the underlying mechanisms. In every case, this was later replaced by a selective theory.' For more discussion, see Piattelli-Palmarini (1986) and Jerne's Nobel Prize address (1985).

Under current formulations, the linguistic genotype, UG, consists of a set of invariant principles and a set of parameters that are set by some linguistic environment, just as certain receptors are 'set' on exposure to a horizontal line. So the environment may be said to 'select' particular values for the parameters of UG. UG must be able to support the acquisition of any human grammar, given the appropriate triggering experience. Of course, UG need not be seen as homogeneous, and may emerge piecemeal, parts of it being available maturationally only at certain stages of a child's development. Grammars must not only be attainable under usual childhood conditions, but also usable for such purposes as speech production and comprehension, appropriately vulnerable for the kinds of aphasias that one finds, and they should provide part of the basis for understanding the developmental stages that children go through. There is no shortage of empirical constraints on hypotheses about (1).

The "logical problem of language acquisition" has provided much of the empirical refinement of (1). Apparent *poverty of stimulus* problems have led grammarians to postulate particular principles and parameters at the level of UG. The stimulus or trigger experience that children have appears to be too poor to determine all aspects of the mature capacities that they typically attain. It is too poor in three distinct ways: (a) the child's experience is finite but the capacity eventually attained ranges over an infinite domain; (b) the experience consists partly of degenerate data which have no effect on the emerging capacity; and, most important, (c) it fails to provide evidence for many principles and generalizations which hold of the mature capacity. Of these three, (a) and (b) have been discussed much more frequently than (c), although (c) is by far the most significant aspect and provides a means for elaborating theories of UG. For discussion, see Chomsky (1965, Ch. 1), Hornstein and Lightfoot (1981, pp. 9–31), Lightfoot (1982, Ch. 2).

2a   NP   →   Spec   N'

    N'   →   (Adj) $\left\{ \begin{array}{c} N' \\ N \end{array} \right\}$ YP

2b   NP   →   NP   YP

    NP   →   Spec   (Adj)   N

Any argument from the poverty of the stimulus makes crucial assumptions about the nature of the triggering experience. To illustrate, I shall briefly rehearse one argument, discussing some material from Baker (1978), which was refined in Hornstein and Lightfoot (1981) and then further in Lightfoot (1982). It has been generally agreed for a long time that linguistic expressions are made up of sub-units and have an internal hierarchy. It is also generally agreed that a grammar (in the sense defined) is not just a list of expressions but is a finite algebraic system which can "generate" an infinite range of expressions. One might imagine, in that case, that English noun phrases have the structure of either (2a) or (2b).

If the phrase structure rules generating noun phrases are those of (2a), a phrase like *the old man from New York* will have the internal structure of (3a); if the rules are those of (2b), the structure will be (3b).

3a

```
              NP
          /        \
       Spec          N'
       the        /      \
              N'           PP
           /      \      from NY
        Adj        N'
        old         |
                    N
                   man
```

3b

```
                      NP
              /                 \
           NP                    PP
        /   |   \              from NY
     Spec  Adj   N
      the  old  man
```

In (3a) *the old man,* for example, is not a single unit, but in (3b) it is. The crucial difference is that the rules of (2a) refer to N', an element intermediate between the nucleus noun and the maximal noun phrase (NP).

Now, it can be shown that any noun phrase that occurs in English, and thus any noun phrase that an English-speaking child will hear, can be generated by both sets of rules. However, linguists believe that something along the lines of (2a) must be correct, or at least preferred to (2b), because (2b) is consistent with certain phenomena which do **not** occur in English and because (2b) fails to account naturally for certain ambiguities. (2b) has no N' node, and therefore provides no straightforward way to distinguish between (4a and 4b), and no ready means to capture the ambiguity of (5a), which may have the meaning of (5b) or (5c). The details of the analysis need not concern us here.[1]

4a. * the student of physics is older than the one of chemistry
4b. the student from NY is older than the one from LA
5a. he wants an old suit but he already has the only one I own
5b. he wants an old suit but he already has the only suit I own
5c. he wants an old suit but he already has the only old suit I own

Here is the problem. It is reasonable to suppose that children might be exposed to any noun phrase that might occur in English, but it is not the case that they are systematically informed that sentences like (4a) are not uttered by most speakers and that (5a) has two possible meanings. In fact, perception of ambiguity is a sophisticated skill which develops late and not uniformly; most ambiguities pass unnoticed and people take the most appropriate of the available meanings. To be sure, children come to know these things and this knowledge is part of the output of the language acquisition process, but it is not part of the input, not part of the "evidence" for the emerging system, and thus not part of the triggering experience. Consequently, if linguists prefer hypothesis (2a) over (2b) on the basis of phenomena like (4) and (5), children have no analogous basis for such a choice if such data are not available to them. In that case they must arrive at (2a) on some other, presumably non-experiential basis. So linguists have postulated genotypical information that phrasal categories have the structure of (6). By (6a) any noun phrase (NP) consists of a Specifier and a N' in some order to be determined by the child's particular linguistic experience, the "trigger" of (1a); similarly a verb phrase, VP, consists of a Specifier and a V' in some order, and likewise the other phrasal categories. By (6b) the N' consists of a nucleus (N or N') and some satellite material in some order (the comma indicates an unordered set).

---

[1](2b) provides only one possible structure for a noun phrase consisting of a nucleus noun followed by a preposition phrase, whereas (2a) provides more than one structure: *student from NY* can (and must) have the structure $_{N'}[_{N'}[_N[student]]_{PP}[fromNY]]$, while *student of physics* is $_{N'}[_N[student]_{PP}[of physics]]$. The process of interpreting the pronoun *one* refers to a preceding N'. *Student* is a N' in (4b), hence a referent for *one*, but not in (4a). In (5a) both *suit* and *old suit* are instances of N' and thus possible referents for *one*, hence the ambiguity. For details and the reasons why *student from NY* and *student of physics* must have different structures, see Lightfoot (1982).

6a. XP $\longrightarrow$ Spec, X'

6b. X' $\longrightarrow \left\{ \begin{array}{c} \text{X'} \\ \text{X} \end{array} \right\}$, Adj, YP

7a. the house
7b. crazy people, students of linguistics

(6a) and (6b) are parameters that are set on exposure to some trigger. The English-speaking child hears phrases like (7a) and, after some development, analyzes them as consisting of two words, one of a closed class and the other of an open class; in the light of this prior knowledge and in the light of the parameter (6a), the child adopts the first rule of (2a). Likewise, exposure to phrases like (7b) suffices to set parameter (6b), such that the second rule of (2a) is adopted. Given the parameters of (6), rules like those of (2b) are never available to children and therefore do not have to be "unlearned" in any sense. Although no "evidence" for the existence of a N' node seems to be available in a child's experience, it is provided by the genotype and therefore occurs in mature grammars (I shall consider an alternative account later).

There is much more to be said about this argument and about its consequences. I have sketched it briefly here in order to demonstrate that any poverty of stimulus argument is based on certain assumptions about the triggering experience. The assumption so far has been that the non-occurrence for many people of (4a) and the ambiguity of (5a) are not part of the trigger, but that data like (7) are. It should be clear that there is a close relationship between the three entities of (1), and a claim made about any one of them usually has consequences for hypotheses about the other two. If the primary linguistic data (PLD) were rich and well-organized, correspondingly less information would be needed in UG, and vice versa. These are not aesthetic swings and roundabouts, and there are clear facts of the matter which limit viable hypotheses.

I shall argue that the trigger consists of a haphazard set of utterances made in an appropriate context and of a type that any child hears frequently. In other words, it consists of robust data and includes no "negative data," information that certain expressions do not occur. I shall flesh this out, making it more precise and more controversial, but first I shall contrast it with some other ideas in the literature.

It is clear that the PLD which trigger the growth of a child's grammar do not include much of what linguists use to choose between hypotheses. To this extent the child is not a "little linguist," constructing her grammar in the way that linguists construct their hypotheses. For example, the PLD do not include well-organized paradigms nor comparable data from other languages. Nor do the PLD include rich information about what does not occur, i.e. negative data.[2] It is true that some zealous parents correct certain aspects of their child's speech and so provide negative data, but this is not the general basis for language development. First, such correction is not provided to all children and there is no reason to suppose that it is an indispensible ingredient for language growth to take place. Second, even when it is provided, it is typically resisted, as many parents will readily attest. McNeill (1966, p. 69) recorded a celebrated illustration of this resistance.

Child: Nobody don't like me
Mother: No, say "nobody likes me."
Child: Nobody don't like me (eight repetitions of this dialogue)
Mother: No, now listen carefully; say "nobody likes me."
Child: Oh, nobody don't likes me.

Third, correction is provided only for a narrow range of errors, usually relating to morphological forms. So, the occasional *taked, goed, the man what we saw*, etc. might be corrected, and McNeill's child on the eighth try perceived only a morphological correction, changing *like* to *likes*. However, not even the most conscientious parents correct deviant uses of the contracted form of verbs like *is* and *will* (8), ... in this case because they do not occur in children's speech.

8a. Jay's taller than Kay's (cf. ... than Kay is)
8b. Jay'll be happier than Kay'll (cf. ... than Kay will)

They also do not correct errors in which anaphors are misused. Matthei (1981) reports that children sometimes interpret sentences like *the pigs said the chickens tickled each other* with *each other* referring to *the pigs*. This misinterpretation is unlikely to be perceived by many adults. Similarly with many other features of children's language. For good discussion, see Baker (1979).

It is sometimes argued that while children are not supplied with negative data directly, they may have access to them indirectly. So Chomsky (1981, p. 9) speculates along these lines:

> if certain structures or rules fail to be exemplified in relatively simple expressions, where they would be expected to be found [my

---

[2] Young children are known to have great difficulty in detecting for themselves the absence of forms, even when confronted with carefully prepared paradigmatic sets of patterns (Sainsbury 1971, 1973).

emphasis-DWL], then a (possibly marked) option is selected excluding
them in the grammar, so that a kind of "negative evidence" can be available even without corrections, adverse reactions, etc.

This is illustrated by the so-called null-subject parameter, whereby
expressions like (9), with a phonetically null subject, occur in Italian,
Spanish and many other languages, but not in English, French, etc.

9a. ho trovato il libro
9b. chi credi che partirà?
9c. * found the book
9d. * who do you think that will leave?

Chomsky, following Rizzi (1982), suggests that if the English-speaking
child picks the wrong setting for this parameter, then failure to hear
sentences like (9c) might be taken as indirect evidence that such sentences are ungrammatical and thus do not occur for some principled
reason. Consequently the child will pick the setting which bars (9c,d).

Two comments on this. First, if children do have indirect access
to negative data, it will have to be specified under what circumstances.
That is, in Chomsky's formulation above, the phrase "in relatively
simple expressions, where they would be expected to be found" will
need to be fleshed out in such a way that it distinguishes cases like (9)
from those like (4) and (8), etc. While one might argue that children
may have indirect access to data like (9c), it is hardly plausible to say
that they have indirect access to (8). For this distinction to be made,
UG would have to be enriched to include analogical notions which have
not yet been hinted at.

Second, so far there are no strong arguments for indirect access
to negative data. Certainly there are plausible alternatives for the
null-subject parameter. One possibility is to claim that the English
setting for this parameter is **unmarked,** i.e. the default case. Thus
Italian and Spanish children need specific evidence to adopt the other
setting, and (9a) is the required evidence.[3] The fact that the Italian
setting of the null-subject parameter seems to be much more common
across languages than the English setting does not entail that it is less

---

[3]The notion of "markedness" has led to much confused discussion. UG
includes a theory of markedness which leads one parameter setting to be
preferred over another and permits "core grammar" to be extended to a
marked periphery (Chomsky 1981, p. 8). So the unmarked parameter
setting is adopted in the absence of contrary evidence, but specific
evidence will be required for a marked setting.

marked, since markedness values do not reflect statistical frequency. In fact, Berwick's Subset Principle (1985) predicts that the Italian setting should be marked. The Subset Principle requires children to "pick the narrowest possible language consistent with evidence seen so far" (p. 237). The Italian setting of the parameter entails a language which is broader than one with the English setting, and therefore the English setting needs to be unmarked (p. 290).

A second possibility is to make the null-subject parameter dependent on some other parameter. It has often been suggested that null subjects occur only in grammars with rich verbal inflection. However, rich inflection seems to be a necessary condition for null subjects, but not sufficient. So German does not have null subjects, although its verbal inflection seems to be as rich as that of Spanish, which does allow null subjects. Consequently, the learning problem remains constant and is unaffected by the richness of inflections. As an alternative, Hyams (1983) related the impossibility of null subjects to the occurrence of expletive pronouns (*it's cold, there's no more*) and she marshalled some interesting evidence in favor of something along these lines by considering the developmental stages that children go through.

Indirect access to negative data may prove to be needed but so far no very plausible case has been made.[4] The notion raises non-trivial problems in defining the contexts in which indirect access is available. Meanwhile plausible solutions for problems which seem to call for indirect access to negative data are suggested by viewing the phenomena in relationship to other parameters and not in isolation. I have mentioned two such parameters here, but there are other suggestions in the literature.

Putting aside further discussion of the possibility of indirect access to certain negative data, one can plausibly argue that the triggering experience, then, is less than what a "little linguist" might encounter and

---

[4]Baker (1979) discusses a transformational movement rule that relates *John gave the book to Alice* and *John gave Alice the book,* which does not generalize to *report* and *say.* The fact that the rule is not entirely general suggests that negative data are needed to establish the limits to the generalization. He went on to show that a lexical relationship is preferable to a movement analysis and circumvents the apparent learnability problem if children are conservative in establishing the lexical properties of verbs, generalizing only within narrowly prescribed limits. For further discussion, see Mazurkewich and White (1984) and Randall (1986).

does not include information about starred sentences and much more
that occurs in a typical issue of the technical journals. Such things are
simply not part of a typical child's linguistic experience. Consequently,
we may persist with the idea that the trigger consists of nothing more
than a haphazard set of utterances in an appropriate context. However,
we can restrict things further: the trigger is something less than the
total linguistic experience. The occasional degenerate data that a child
hears and idiosyncratic forms do not necessarily trigger some device in
the emergent grammar which has the effect of generating those forms.
So, for example, a form like (10a) might occur in a child's experience
without triggering an unusual form of subject-verb agreement. Similarly a New York child might hear (10b) without having *y'all* triggered
as a word in his or her grammar.

10a. the person who runs the stores never treat people well
10b. y'all have a good time in South Carolina

A child might even be exposed to significant quantities of linguistic
material which do not act as a trigger. So, if a house-guest speaks an
unusual form of English, perhaps with different regional forms or the
forms of somebody who has learned English imperfectly as a second
language, this normally has no noticeable effect on a child's linguistic
development. Even children of heavily accented immigrant parents
perpetuate few non-standard aspects of their parents' speech.

I take it that this is intuitively fairly obvious, and shows that there
is little to be learned about the trigger experience from simply tape-
recording everything uttered within a child's hearing (cf. Wells, et al.
1981). More can be learned from the historical changes that languages
undergo. It is well-known that certain kinds of syntactic patterns become obsolete in certain speech communities at certain times. What
this means is that sometimes speakers hear a form which does not trigger some grammatical device which permits it to be generated and thus
to occur in their mature speech. The conditions under which this happens cast some light on the nature of the trigger, as I shall discuss in
the next section.

So then, we may now claim that the trigger experience is some
subset of a child's total linguistic experience. But where exactly are
the limits? This is often a crucial question in grammatical analyses,
but it is rare to see alternatives discussed. Consider again the example
of the structure of noun phrases. I argued above that any noun phrase
that an English-speaking child could hear would be consistent with
both the rules of (2a) and (2b). I also claimed that the data which

lead grammarians to prefer (2a) to (2b) are generally not available to children and therefore that the information which eliminates (2b) must come from some other, presumably genetic source. However, one could look at things somewhat differently. The real difference between (2a) and (2b) is the existence of a N' node in the rules of (2a). The existence of this node is required by the UG parameter of (6b) and, on that account, does not have to be derived from relevant experience. In that case, we might ask if there is anything in a child's experience which would require postulating a N' node, and one can indeed imagine evidence which would force the child to establish such a node.

English speakers use the indefinite pronoun *one* to refer back to a N', as noted earlier, and the fact that it refers to a N', something intermediate between a NP and a N, might in fact be learnable. A sentence like (11a) would not be a sufficient basis for learning this because, regardless of whether Heidi actually has a big or small cup, the sentence could always be interpreted as specifying only that Heidi had some cup regardless of size (with *one* referring only to the N *cup*). Sentence (11b), however, would suffice if uttered in a situation where Heidi has a cup that is some color other than blue; only the interpretation with *one* representing *blue cup* would be consistent with the facts. In that case a child might **learn** correctly that *one* must refer to something larger than a N, namely a N'.[5]

11a. Kirsten has a big cup, and Heidi has one too
11b. Kirsten has a blue cup, but Heidi doesn't have one

We now have two alternative accounts: the existence of N' might be derived from a property of UG or it might be triggered by the scenario just sketched. My hunch was and remains that this scenario is too exotic and contrived to be part of every child's experience, and therefore that postulating (6) at UG is more plausible. But this hunch may be wrong. It is certainly falsifiable. If parameters like (6) exist at UG, then strong claims are made about the possible degree of variability that will be found in the languages of the world: in languages where this kind of structural configuration is relevant (which may or may not be **all** languages), there will be essentially four NP types (12). Type (12a) is represented by English, French, etc. and type (12b) seems to be mainfested in Basque, Burmese, Burushaski, Chibcha, Japanese, Kannada, and Turkish (see Greenberg 1966, p. 20). Types (12c,d)

---

[5]Sentences like *Kirsten has* NP[a blue cup] *and Heidi has a red one* show that *one* must also refer to something smaller than a NP.

are more problematic because I know of no carefully studied grammar which manifests them. Greenberg (1966) and Hawkins (1979) discuss several languages in which demonstratives follow the head noun and which therefore might be of type (12c or d), but they do not distinguish between demonstratives which have the syntax of adjectives (as in Latin) and those which manifest Spec (as in English). If it should turn out that types (12c,d) do not occur, then parameter (6a) will be tightened to allow only the Spec–N' order.

12a. $_{NP}$[Spec $_{N'}$[nucleus satellite]]
12b. $_{NP}$[Spec $_{N'}$[satellite nucleus]]
12c. $_{NP}$[$_{N'}$[nucleus satellite] Spec]
12d. $_{NP}$[$_{N'}$[satellite nucleus] Spec]

Also, parameters like (6) suggest that one will find developmental stages corresponding to the fixing by a child of the two parameters. Lightfoot (1982, p. 179f), building on work by Klima and Bellugi (1966) and Roeper (1979), argues that this is indeed the case. Children seem to acquire noun phrase structures in four identifiable stages. Examples (13a,b) list some noun phrases occurring in the first two stages.

13	a. car	b. a coat	that Adam
	baby	a celery	more coffee
	wa-wa (water)	a Becky	two socks
	mama	a hands	big foot
	hands	my mommy	

All children go through the four stages at some point, although the ages may vary. Most children utter the stage 2 forms between one and two years. At stage 3 there is more sophistication.

14	mama	my doll	a blue flower
	cracker	your cracker	a nice cap
	doll		a your horse
	spoon		that a horse
			that a blue horse
			your blue cap

At stage 4 the mature system emerges, which normally remains more or less constant for the rest of the child's lifetime. But consider (15), some forms that never occur in children's speech.

15	* blue a flower	* a that blue flower	* flower a
	* nice a cup	* blue a that	* house that a
	* my a pencil	* that a	
	* a that house	* a my	
		* my a	

Recall the parameters for noun phrases developed earlier. These were hypotheses about how NP structure could vary from grammar to grammar. At stage 1 these principles are irrelevant, because the child has only one-word structures. Other cognitive capacities are relevant, such as the conceptual system that involves properties and conditions of reference, knowledge and belief about the world, conditions of appropriate use, and so on. These play a role in explaining why *mama* and *cup* are more likely than *photosynthesis, quark* or *grammar* to be among the earliest words in a child's speech.

At stage 2 the child seems to have fixed the first parameter and determined that the order is Spec N': all specifiers appear at the front of the noun phrase. The occurrence of phrases like *a Becky, a hands* suggests that children cannot distinguish at this stage definite and indefinite articles, and that they do not know that *a* is singular. There is no evidence that the child can distinguish subtypes of specifiers (articles, possessives, numerals, demonstratives), but they all occur one at a time in front of a noun.

By stage 3, children discriminate some kinds of specifiers and establish some more of the relative orders. In fact, the child knows that all specifiers precede adjectives, which in turn precede nouns, and that specifiers are optional, while the noun is obligatory. The stage 3 grammar differs from the mature system in that the child does not yet know that an article may not co-occur with a demonstrative or with a possessive like *your*. This suggests that the child now has the PS rules NP → Spec N', N'→ (Adj) N, but that it takes a little longer to determine the status of a demonstrative and whether a form like *your* is a specifier or an adjective. After all, in other languages demonstratives and possessives are often adjectives instead of specifiers (e.g. Italian *la sua machina* 'his bicycle').

Consequently, there is reason to believe that postulating the parameters of (6) at UG is more plausible than claiming that the existence of N' is **learned** on the basis of exposure to sentences like (11b) uttered in the relevant context. But the important point is that alternatives like this need to be sketched and evaluated, and that grammarians should be paying more attention to their assumptions about the nature of the triggering experience required to set the parameters they hypothesize.

So I persist with the idea that the trigger is a subset of a child's experience, and that it probably does not include exotic events like the one sketched above in the context of (11b). The trigger consists only of robust data which can be analyzed consistent with genotypical principles and already fixed parameters of the child's grammar. The question remains of how small the subset is.

There is a theory, advanced by Snow (1977, etc.) and others, that the crucial input for language growth to take place is very small, a specially structured form of speech transmitted through mothers and caretakers. This "motherese" is supposed to provide a set of patterns which are generalized by children on an inductive basis. This view was held fairly widely for a while.

There are at least four reasons why this kind of pattern generalization is not the answer to how children acquire speech. First, although children no doubt register only part of their linguistic environment, there is no way of knowing quite what any individual child registers. Therefore factual basis is lacking for the claim that children register only what is filtered for them through parents' deliberately simplified speech. Children have access to more than this, including defective utterances. Second, even supposing that they register only perfectly well-formed expressions, that would not be enough to show that the child has a sufficient inductive base for language acquisition. Recall that the child's stimulus is "deficient" in three distinct ways (p. 3 above); the motherese hypothesis would circumvent only the degeneracy problem (b) but leaves untouched problems (a) and the far more important (c), the absence of evidence in PLD for certain partial generalizations. The poverty of stimulus problems still hold and the child would need to know that the contractability of the first *is* in (8) could not be extended to the second *is*. One wants to know why quite ordinary inductive generalizations like this are in fact not made; the so-called motherese does not show where inductive generalizations must stop. Third, if the child registered only the simplified and well)formed sentences of motherese, the problem of language learning would be **more** difficult because the child's information would be more limited. Fourth, careful studies of parents' speech to children (like Newport, Gleitman and Gleitman 1977) show that an unusually high proportion consists of questions and imperatives; simple declarative sentences are much rarer than in ordinary speech. This suggests that there is very little correlation between the way the child's language emerges and what parents do in their speech directed at children. Thus, the existence of motherese in no way eliminates the need for a genetic basis to language acquisition. The child is primarily responsible for the acquisition process, not parents or older playmates. (For good discussion of this topic, see Wexler and Culicover 1980, pp. 66–78).

Furthermore, while it is by no means clear exactly what this motherese consists of, the general phenomenon is not uniform and does not occur in all households or cultures. Even where motherese is not practised, children nonetheless attain a normal linguistic capacity. This suggests that the child's trigger experience does not need to be limited artificially along the lines of motherese.

For the last fifteen years grammarians have been seeking to develop locality restrictions, such that grammatical processes only affect elements which are not too far apart. This work suggests that in general grammatical processes affect only items which are clause-mates or where the item in a lower clause is, loosely, at the front of that clause. Locality restrictions are formulated somewhat differently at different stages of the development of UG and by different authors. The details of various locality restrictions need not concern us immediately, but they do raise the following question: if grammatical processes are generally limited to clause-mates or at most to items of which one is at the front of an embedded clause, why should children need to hear more than a single clause (plus the front of a lower clause) in order to hear the effects of all possible grammatical processes in their language? In fact, a good case can be made that everything can be learned from essentially main clauses (degree-0 learnability); see Lightfoot (1987).

These are some of the issues which arise in the context of the development of language in an individual, ontogenetic development. The mature linguistic capacity emerges in a child as the environment sets the parameters which are prescribed genetically. For a normal capacity to emerge, the child needs access only to simple, robust expressions. With this perspective, we can now consider what some of the boundary conditions would be for models of phylogenetic change, which account for how a language might change from one generation to the next, how Old English became Middle English, how Latin became Italian or Spanish or French.

3. Phylogeny: Language Change

Anybody who has attended a performance of *Macbeth* or read the King James version of the Bible knows that English has changed over the last 400 years. Shakespeare's sentence structures were not like today's English, although most of them are not difficult to understand. Difficulties do arise, however, if one goes back 200 years more to Chaucer's *Canterbury Tales,* where the language is a good deal less familiar. Going back much further to *Beowulf,* and one may as well be reading a foreign language.

Not only does sentence structure change in the course of time, but so do the meanings of words, their form, and their pronunciation.

Consider the first four lines of Shakespeare's Sonnet XI:

> As fast as thou shalt wane, so fast thou grow'st
> In one of thine, from that which thou departest;
> and that fresh blood which youngly thou bestow'st
> thou mayst call thine when thou from youth convertest.

A modern speaker would not use *convert* in the sense of line 4. There are unfamiliar word forms: *shalt, growth'st, thine. Depart* can no longer be used as a transitive verb. Lines 2 and 4 do not rhyme today, but they did rhyme for Shakespeare, who pronounced *convert* as if the *e* were an *a* (as in the modern British pronunciation of *clerk, Derby*). So, more difficulties would arise for a modern audience if actors tried to imitate Shakespeare's pronunciation. Those difficulties would be greater if a time machine enabled us to hear Chaucer reading his own poetry: the vowel of *ripe* would be pronounced like the modern *reap,* and *rout* like modern *root,* and the language might sound about as alien to our ears as Dutch does today.[6]

It is normal, then, that languages change over time in many different ways. Over a large enough time span, the changes may result in the language being quite different from what it once was. So Old English is quite different from Modern English. Italian has descended more or less directly from Latin but the speech of Sophia Loren would be incomprehensible to a revived Cicero.

There is an old explanation for why there are so many languages and dialects in the world and why they are constantly changing. It says that people once tried to build a tower high enough to reach heaven, the Tower of Babel. For this act of hubris God punished them, making them speak different languages so that they would no longer be able to understand each other. Whether we believe this or not, it is clear that languages do not differ from each other or change historically in entirely arbitrary ways. Perhaps we can build a better explanation, without being guilty of a new form of hubris.

We aim to discover the invariant properties of grammars, where a grammar characterizes the subconscious linguistic knowledge of a mature speaker, and to specify the genetically determined properties that permit children to master their languages. We assume that many properties of the mental genotype are invariant from person to person, whether people living in America or China today or people living in

---

[6] Looking at a written page involves other difficulties because of changing spelling conventions and even different symbols, like þ for *th.*

modern or medieval France. Of course, going back far enough in time might reveal significantly different human genotypes, but these are unlikely to be revealed for a tiny period of some 3,000 years, the period for which evidence about language has been found. No records exist to support claims about people's grammars of more than 3,000 years ago. Therefore language change must be viewed as reflecting different grammars, all attained in the usual way on the basis of a more or less common genetic inheritance. Shakespeare's internalized grammar was different from mine because it generated a different set of structures, but presumably it developed in him in accordance with the same genetic program that enabled Ezra Pound and Tennessee Williams to develop their quite different linguistic capacities.

Although particular grammars may change from one generation to the next, all grammars of the last few thousand years can be assumed to accord with the theory of grammar and with our common genetic inheritance and to be attainable by a child on the basis of exposure only to the usual linguistic environment. Looking at the point where certain changes take place may inform us about the limits to attainable grammars, about when the linguistic environment changes in such a way as to trigger a different kind of grammar. I shall show how this might work out in practice, illustrating with changes affecting the syntactic component, but one can take an analogous approach to phonological and semantic changes.

A good theory of grammar illuminates the nature of historical change but one must be careful not to demand too much. A theory of grammar should not seek to explain all the changes that a language might undergo, because the history of a language is not fully determined by the properties it shows at some arbitrary starting point and by the properties of the mental genotype. Many changes are due to other things, some of which can be regarded as chance or at least nongrammatical factors.

A moment's reflection will show that this must be the case, because a language may split in two and then pursue different courses, diverging more and more. Latin, for example, developed into French, Rumanian, Italian, and several other identifiable languages. Most European and several languages spoken in India descend ultimately from a language called Proto-Indo-European, for which there are no direct records. The parent language is usually supposed to have had an underlying word order of subject-object-verb (SOV); its descendants have quite different orders: Hindi has retained SOV, English has developed SVO, Welsh and Irish have VSO. Given the possibility of divergent development, historical changes cannot be fully determined by the properties of Proto-Indo-European and the (invariant) demands of the theory of grammar, our genetic endowment. The theory of grammar cannot prescribe a universal path for languages to slide along at various rates, acquiring properties in a predestined order; it cannot prophesy which changes a language will undergo in the future, because it cannot predict which chance factors will operate or when.

Nonetheless languages do not change in completely arbitrary ways; many changes recur in one language after another. Despite the role of nongrammatical factors and chance, some changes and the manner in which they arise can be explained and occur as a matter of necessity. The interaction of chance and necessity can be seen in the much studied phenomenon of word order change.

Consider a language changing from SOV to SVO order. English is an example, because the grammar of Old English had a phrase structure rule V'→ NP V, generating object-verb order, and Modern English has V'→ V NP. So this phrase structure parameter came to be fixed differently. How could this change have taken place?

First, while the sentences of Old English are very different from those of Modern English, differences are much smaller over a shorter time span. In the speech of two adjacent generations, some parents and their children, the difference between the sentences uttered is usually quite small. This is not surprising because the speech of the immediately preceding generation, particularly that of the parents, usually provides a significant part of the linguistic environment that triggers the development of the child's grammar.

For any usual parent and child, the output of their grammars will be quite similar.[7] This follows from the way in which children develop grammars, stimulated by their linguistic environment, and it imposes a tight limit on the class of possible historical changes. For example, no grammar with phrase structure rules generating SOV order could undergo a re-analysis of its rules to generate initial SVO structures unless there were already some SVO sentences in the language. This fact about word order changes is explained by the way in which children

---

[7]It does not follow from this that their grammars will be similar, because there is no one-to-one relation between difference of output and difference of internal grammar. Two wildly different grammars might yield fairly similar outputs; conversely two similar-looking grammars might yield wildly different outputs.

acquire grammars. A consistently SOV linguistic environment would not trigger a grammar with basic SVO order, that is, with the V'→ V NP phrase structure rule. Therefore an order change from SOV to SVO cannot take place without warning, suddenly and across the board. However, change is not always gradual.

A speaker whose grammar has rules generating an initial object-verb order may nonetheless utter sentences with verb-object order. In fact, it seems that all underlying SOV grammars "leak," allowing some surface instances of orders other than SOV. This is due to limitations of human processing abilities, to the fact that multiple center-embeddings strain short-term memory capacity. A SOV language incurs the danger of center-embedding with object complements, because the grammar would generate initial structures like (16). Consequently a SOV grammar will have some device eliminating these structures—for reasons of necessity relating to short)term memory capacity. A transformational rule converting SOV to SVO would suffice: (16) would be transformed into the right-branching and harmless (17), where there is no center-embedding.

16

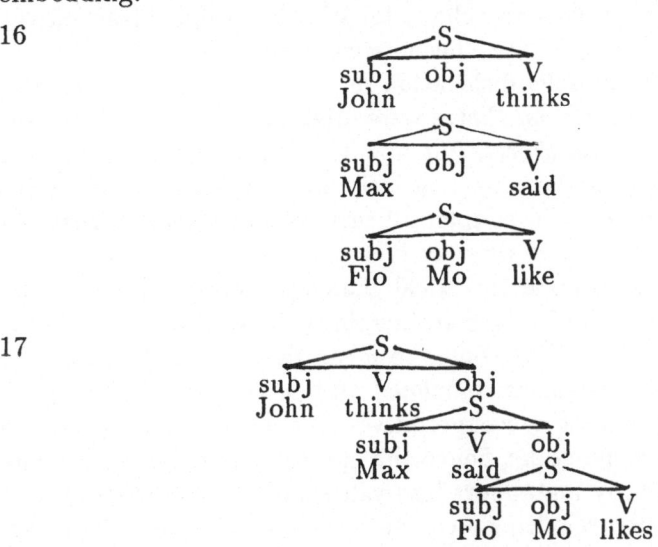

17

Many SOV languages show SVO order when the object consists of a sentence. No doubt there are other solutions, so presumably there was nothing necessary about the fact tht this particular solution was adopted by Classical Greek, Old English, Dutch and German.

It is fact of biological necessity that languages always have devices to draw attention to parts of sentences, and people may speak more expressively by adopting a novel or unusual construction, perhaps a new word order. When they first occur, these novel forms may not be part of the output of a grammar but may be quite irregular, specially learned accretions in the way that formulaic expressions like "Good morning," "Hi," "Wow," are part of my language but quite idiosyncratic and not a function of any general rule.[8] Although there are limits, one cannot forecast which novel forms will be introduced or when; least of all can one forecast which novelties will catch on and be perpetuated. Chance or nongrammatical factors are at work. At a later stage the forms originally introduced as novelties may become "grammaticalized" and have a general, predictable and rule-governed distribution. Grammars at that stage would have a transformation or some other device that has the effect of permitting them to be generated.

Dislocation sentences fall under this rubric: *Mingus, I heard him* and *he played cool, Miles.* These forms, they are still regarded as novel in English and as having a distinct stylistic force, focusing attention on the noun phrase; they are common in Yorkshire dialects, with British sports commentators and in the speech of many Jewish Americans. However, such expressive forms characteristically become bleached and lose their novelty value as they become commonly used. This can be illustrated with the parallel dislocation sentences in French: *Pierre, je le connais* and *je le connais, Pierre* were originally stylistically marked alternants of *je connais Pierre,* but now they have lost much of their special force and have become relatively unremarkable construction types, to the point that in simple, affirmative main clauses they are the norm and the former *je connais Pierre* is vanishingly rare.

This process is familiar in lexical change, where, to the constant dismay of the purists, adjectives are regularly "devalued" by a kind of linguistic inflation: *excellent* comes to mean merely 'good', *enormous* to mean 'big', and *fantastic, fabulous,* and so on lose their original force. As this happens, so new superlatives must be invented to describe the end point on some scale: hence the currently popular *ginormous, fantabulous.* Similarly metaphors lose value and become standardized through frequent use, requiring a constant effort on the part of speakers to find new forms with the old surprise value.

---

[8] The existence of such forms is one reason why a person's grammar does not define all of the expressions of his or her language, and why there may not be a mechanical device that can generate all these expressions. These forms are presumably learned, that is, shaped in direct response to experience, perhaps even in adulthood.

So in syntax new constructions are introduced, which by their unusual shape have a novelty value and are used for stylistic effect. The special stylistic effect slowly becomes bleached out and the constructions lose their particular force, become incorporated into the normal grammatical processes and thereby require speakers to draw on their creative powers once more to find another new pattern to carry the desired stylistic effect. So sentences such as *Mingus, I heard him* seem to be fairly recent innovations and have a special focusing effect in most dialects; in other dialects, notably those of North Americans with a Yiddish background, the construction has already become bleached of its special effect, like the dislocation sentences of French. This is an important kind of change in syntax.

Again, although languages necessarily have some means for expressiveness, the particular means varies (within limits) from language to language. Given a grammar generating object-verb order, speakers may come to develop a transformation allowing verb-object order in some contexts, perhaps for reasons of focusing.

That transformation will not be "structure preserving," since the output (verb-object) is not generated directly by the phrase structure rules; therefore it can apply only in main clauses. This is a matter of necessity, dictated by human genetic endowment (see Lightfoot 1982 for discussion). This entails that any innovation that is a function of a non-structure-preserving transformation enters a language first through main clauses, only later percolating through to affect structures in embedded clauses. The manner of these changes is explained by a theory entailing that non-structure-preserving rules affect only main clauses.

Numerous examples can be cited of changes progressing in this fashion. The SOV–to–SVO change affected English first in main clauses and later in subordinate clauses. Modern Dutch and German are like early Middle English in having a more recent SVO order in main clauses and an older SOV order elsewhere. Basque now seems to be undergoing a similar change in a similar way. Such changes are entirely consistent with current theories, unlike, say, a change affecting first subordinate clauses and then spreading to all other clause types. Such a change could not be interpreted in this theory, because it is not possible to have a movement rule that affects only subordinate clauses. The theory therefore explains why changes do not percolate up historically from subordinate clauses to main clauses, whereas the reverse direction is quite common.

Linguists studying change have traditionally noted that main clauses are the most progressive environment; innovations are first intro-

duced there and spread later into other environments. In fact, the
"structure-preservation" idea yields finer predictions than the traditional account. It entails that rules moving major categories enter the
grammar as processes affecting main clauses before subordinates; it
does not follow that a morphological change, say, should affect verbs
in main clauses before subordinates. In other words, only certain kinds
of change will percolate from main to subordinate clauses. This finer
prediction is more consistent with the facts.

If the earlier appeal to perceptual modes is correct, all SOV grammars have in them the seeds for a change to underlying SVO order.
This is becasue they typically show SVO order when the object consists of a sentence, as noted. The language may also develop a focusing
process and perhaps other devices that yield a SVO order. If the SVO
forms proliferate sufficiently, at some point the linguistic environment
will trigger in children an **underlying** SVO order as part of the most
readily attainable grammar; that is, the phrase structure rules will
come to generate verb-object order instead of the earlier object-verb.
This kind of change represents a new parameter setting; it arises as a
matter of necessity dictated by the theory of parameters. Given the earlier changes, the linguistic environment has become such that the best
grammar must contain a phrase structure rule V'→ V NP. Presumably
an underlying SOV grammar with a number of transformational processes yielding SVO order was unattainable under these circumstances,
being too complex and opaque; a theory would explain the change if
it could specify that this degree of complexity makes the grammar less
readily attainable than one with the phrase structure parameter set
differently.

As an indication that a parameter is set differently and that a
change has taken place in the phrase structure rules, one would look
for a variety of simultaneous changes in the occurring sentences, most
notably a significant increase in the occurrences of the innovating order
in subordinate clauses. There is good evidence that English underwent
just such a re-analysis in the thirteenth century, whereby SVO became
the basic order.

There is also good evidence that this change in phrase structure
rules helped to provoke several further changes, such as a change in
the meaning of the verb *like*. This is worth considering in some detail.
In earlier forms of English sentences like *the king likes the queen* used
to occur with the same meaning that they have today but with *the
king* being analyzed as the object of the verb *like* and *the queen* as
the subject. So a Middle English speaker might also have heard (18),

where the post-verbal noun phrase is clearly the subject. It is plausible to analyse such forms as in (19), where the post-verbal subject is linked to the usual subject position, which is empty (and perhaps the residue of movement). If this is the correct syntactic analysis, then *like* must have meant "cause pleasure for," unlike in Modern English where it always means "derive pleasure from." There were some forty or fifty verbs which could occur in such a syntactic context: other examples were *repent, rue, ail,* etc.

18a. him likes the queen
18b. the king like the pears

19 $_{NP}[e_i]_{VP}[_{V'}[$him likes$]_{NP}[$the queen$_i]]$

Forms like (18) died out during the Middle English period; that is, some speakers must have heard them but did not reproduce them. For those speakers such forms were part of their linguistic environment but were not part of the trigger experience; they did not trigger a grammatical device which permitted their generation, and we know why. Notice that the structure (19) contains a V' with object-verb order. This order was characteristic of early English, just as it is characteristic of modern Dutch and German. Given the phrase structure convention of (6b), speakers of early English had the satellite preceding the nucleus while later English speakers fixed the order differently. My point here is that as the order of elements within the V' changed, so forms like (18) were no longer part of the trigger although they were heard by children. When the V' parameter was fixed differently, a form like (18a) would have a structure like (20).

20 $_{NP}[e_i]\ _{VP}[_{V'}[$likes him$]\ [$the queen$_i]]$

The problem is that there is no ready way for *him* to move to a pre-verbal position without violating a principle of grammar. There is only one available NP position for *him* to move to (since there is no evidence that Middle English allowed a pre-verbal clitic NP), and, if it moves to the empty subject position, the indexed empty element would no longer be visible and there would be no way of interpreting *the queen* as the subject with which the verb must agree in number and person. Furthermore, the derivation would violate the theta criterion of Chomsky (1981), because there would be no one-to-one relationship between noun phrases and theta positions (*him* would be associated with two positions and *the queen* with none). Consequently, a form like (18a) could not be derived from a structure like (20).

Nonetheless, children at the relevant stage presumably heard sentences like *the king likes the queen,* uttered in a context in which it was clear that the king was happy and that the queen was the reason for his happiness. Since an analysis like (20) was not available, a ready alternative was adopted where no NP was moved and the pre-verbal NP was interpreted as the subject (21). Forms like (18) were heard but not reproduced; they were replaced by (21b,c).

21a. $_{NP}$[the king] $_{VP}[_{V'}$[likes $_{NP}$[the queen]]]]
21b. $_{NP}$ [he]$_{VP}[_{V'}$[likes $_{NP}$[the queen]]]]
21c. $_{NP}$ [he]$_{VP}[_{V'}$[likes $_{NP}$[the pears]]]]

Under this syntactic analysis, which was forced for the reasons given, the perceived meaning of the sentence entailed that *like* could only be interpreted with a different meaning, the modern "derive pleasure from." Consequently the old meaning of *like* and forms like (18) were not part of the trigger experience and died out of the language, and the reason was that they could not be interpreted consistent with the context in which they were uttered and consistent with a V' containing a nucleus preceding a satellite ... assuming that other aspects of the grammar were triggered as in the immediately preceding generation of speakers.[9]

With this view of change, parallel changes affecting several languages independently are not surprising. Change is a function of chance and necessity: it is sometimes a matter of chance (or at least due to nongrammatical factors) that the linguistic environment should change in a particular way, perhaps incorporating a new kind of expression for a focusing effect or an expression borrowed from a neighboring language. It is a matter of biological necessity that the grammar should be readily attainable, that surface strings should be processible with

---

[9]For example, assuming that the morphological case system had already become dysfunctional. It is quite possible that a different re-analysis would have taken place if the case system had been rich enough and robust enough to force children to interpret the pre-verbal NP as a direct object. This change in the use of *like,* etc. has given rise to much recent discussion: see Allen (1986), Anderson (1986), Elmer (1981), Fischer and van der Leek (1983), Warner (1983). Alternative analyses are possible. For example, instead of (19), the structure of (18a) might contain a topic: $_{TOPIC}$[him$_i$]$_{S'}$[Op$_{i}$$_S$[e$_j$ e$_i$ likes $_{NP}$[the queen$_j$]]]]. In that case the re-analysis might be keyed not to the change in word order but to the obsolescence of empty subjects like *e.*

minimal perceptual difficulty, and that generations should maintain mutual comprehensibility. Such necessities force re-analyses at certain points, encourage certain kinds of rules, and restrict the possibilities for change in any given grammar. In this way the partial similarity of developments in French, English and Lithuanian is explained.

The fact that different languages may undergo parallel changes independently will not surprise a biologist. Evolutionists distinguish two reasons for a certain feature being shared by two or more species: that similarity may be due to common genetic ancestry, being *homologous.* Alternatively, the similarity may be due to a common function but arising independently in each species, being *analogous.* So the wings of birds and insects are analogous features, since they developed independently and the common ancestor did not have wings (see Gould 1978, pp. 254ff).

Since UG specifies what constitutes an attainable grammar, it sheds light on those historical changes where a re-analysis has occurred because part of the old grammar is no longer attainable. Conversely, light is shed on the proper formulation of the parameters of UG when one discovers the point at which such a re-analysis occurs, the point at which the linguistic environment (the child's trigger experience) has changed in such a way that certain aspects of the old grammar become unattainable. We have seen how a phrase structure rule V'→ NP V can become unattainable; likewise the old meaning of *like* in Middle English.

So change is viewed as progressing as a function of chance and necessity, just as Monod (1972) viewed change in genetic structure. Changes may be necessitated by UG; such changes are therefore explained by UG. Conversely, noting the point at which parameters are set differently teaches something about the limits to attainable grammars. What should be looked for as evidence of new parameter setting?

To see what is involved in a new parameter setting, consider the history of the English modals (discussed in detail in Lightfoot 1979). There are two stages to this story. In Old English *can, could, may, might, must, shall, should, will, would, do, did* behaved exactly like normal verbs; there was no reason to assign them to a separate modal category. As a result of changes during the course of Middle English, these items became a distinct subclass of verbs, distinguished by various morphological properties. It was now no longer clear that they were verbs. If they were verbs, they had several exceptional features. Put differently, their verbal nature was opaque, harder to figure out, less readily attainable. In the sixteenth century a radical re-analysis took place whereby a new modal category was introduced and some rules were formulated differently. This suggests that the triggering experience to which children were exposed had changed in such a way that it no longer triggered the grammars which earlier generations had attained. A good theory of UG should be able to characterize why this degree of exceptionality is unattainable by a child.

The earlier grammar was along the lines of (22), where *can, could,* etc. were instances of V, as in modern Dutch and German. The later grammar (23) contained a new phrase structure rule and a new category, Modal (perfective and progressive markers were specifiers of V' at all stages).

22　S → NP INFL VP

　　INFL → Tense

　　VP → Spec　V'

　　V'→ V $\left\{ \begin{array}{c} XP \\ S' \end{array} \right\}$

　　Spec V' → Perfective Progressive

23　S → NP INFL VP

　　INFL → Tense Modal

　　VP → Spec V'

　　V' → V $\left\{ \begin{array}{c} XP \\ S' \end{array} \right\}$

　　Spec V' → Perfective Progressive

　　Modal → *can, could, may, might, do,* ...

This grammatical re-analysis was preceded by a number of morphological changes which had the effect of gradually making verbs like *can, could, may, must,* etc. look quite unlike other verbs:

　　i. The antecedents of most of the modals (*sculan, magan, motan, agan* and *durran*) belonged to an inflectional class known as "preterite-presents." These were strong verbs where the preterite forms had taken on present meaning in pre-Germanic and for which new weak preterites had been made. Other members of this class were *witan* 'to know', *dugan* 'to be of value', *unnan* 'to grant', *purfan* 'to need', *munan* 'to think, remember', *benugan* 'to suffice'. The notable thing about this class is that the third person singular was *sceal, man, ann, dearr,* etc. i.e., it did not have the usual *-ep* ending, the antecedent of the modern *-s.* Thus the modals are conservative; there never was a -s ending for these verbs. However, several verbs either dropped out of the language or out of this inflectional class. The crucial effect was that the antecedents of the modals (including the now uncommon *mun*), the only surviving members, became an identifiable class of verbs, with the unique characteristic that they did not have a fricative suffix for the third person singular.

ii. With the loss of the subjunctive mood, the relation between the present and past tenses of the modals became non-temporal in certain senses. The preterite-presents shared with the weak verbs the feature that there was no phonological distinction between the preterite indicative and subjunctive except in the second person singular. Hence modern *should* corresponds phonologically to an old preterite indicative or subjunctive, and now has some old (non-past) subjunctive meanings, as in *he should do it tomorrow*. This means that the *shall/should, may/might* distinctions are not based simply on tense, unlike, say, the *open/opened* distinction.

As a result of *gradual* morphological changes of this type, the verbs *can, might,* etc. looked less and less like other verbs and were increasingly likely to be analyzed as a different category. We know that they were re-analyzed because several changes that took place together around 1525–1550 all reflect the introduction of the new Modal category: a parameter was set differently such that INFL came to contain a Modal category (23). Under the earlier grammar, where *shall, can,* and so on were instances of verbs, they could occur together side by side as in (24a), just like other verbs (compare *I want to begin now*); they could occur with a progressive marker (24b) or a perfective marker (24c); they could occur in infinitival form (24d). In all of these respects *can, may, must,* etc. behaved like normal verbs.

24a. I shall can do it.
24b. I am canning do it.
24c. I have could do it.
24d. I want to can do it.

Grammar (22) generates the sentences of (24). Such sentences occur in the writings of Thomas More and earlier but have not occurred since the sixteenth century, and a grammar like (23) does not generate them: the rules of (23) clearly allow only one Modal per verb and therefore cannot generate (24a). Under the rules of (23) a Modal cannot occur

to the right of the aspectual marker *have* or *be,* as in (24b,c), or in
an infinitive (because in a nonfinite clause INFL is manifested as *to*
instead of Tense and Modal). Sentences like (24) ceased to occur at
the same time, which is consistent with saying that people used to have
grammars like (22) but came to have grammars like (23).

In a grammar like (22), which assigns *could* and *leave* to the same
category, V, transformational rules could not distinguish them. So neg-
atives might be found to the right of the first V, regardless of whether
it was *could* or *left* (25), and questions like (26) occurred, where the
first verb has been moved to the front of the clause, again regardless of
whether it is *could* or *left.*

25a. John could not leave.
25b. John left not.

26a. could John leave?
26b. left John?

In a grammar distinguishing Modals from verbs, such rules would
be expected to affect one or other class of items but not both. In
fact the (25b) and (26b) forms were replaced with expressions with *do,*
which suggests that only modals could receive a negative or be inverted.
The change was a consequence of the new modal/verb distinction. I
shall return to this in a moment.

These changes occurred simultaneously; they all follow from saying that the later grammar distinguished modal and verbal categories,
whereas the earlier grammar did not. The fact that the surface changes
were simultaneous suggests that in the grammar there was just one
change with various surface manifestations.[10] In general, when such
clusters of simultaneous changes occur, it may be possible to explain
their simultaneity by attributing them to a single change in the abstract grammar. Conversely, if one has a hunch that there is a major
difference between the grammars of, say, Old and Modern English, one
looks for a cluster of simultaneous changes that follow from saying that
the one grammar was replaced by another, that some parameter of
grammar is set differently.

---

[10]So the simultaneity of the changes is explained by their singularity.
An alternative analysis would say that the loss of (24a-d), (25b), and
(26b) were each due to different factors, to independent changes in
the grammar, in which case their simultaneity would be a remarkable
accident.

Saying that one grammar replaced another idealizes away from variation among individuals or even dialects. All speakers of Old English could utter sentences like (24), and no speakers uttered them after the time of Thomas More. Somewhere in between these two stages presumably some but not other people uttered them. Analysts depend on the available texts and we can know only the date of the last attested example of some construction type that still survives in the written literature.[11] Of the spoken language at that time we know virtually nothing. Sometimes we can refine our claims, specifying that the sentence died out in one dialect before another, but for the most part analyses must be conducted at a fairly gross level, abstracting away from such details. Grammars are triggered by linguistic environments, so the linguistic environment of Thomas More, when he was a child, triggered quite a different grammar (22) from that triggered in the next generation (23); that is what it means to say that one grammar replaced another. In no case can we specify *exactly* what the linguistic environment was for any individual and compare it with the environment available for that individual's child, who developed quite a different grammar. But then, we cannot do that for an individual child growing up in London today. So if the idealization of a linguistic community is appropriate for the analysis of Modern English it is presumably also appropriate for Old English.

Let us return to the new negative and inversion sentences, which I said were due to the new Modal/Verb distinction. When the reanalysis took place, there was a discrepancy between experience and production. Some children heard the old forms (*left John?*, etc.) but did not accommodate them; so, these forms were not part of the triggering experience, despite being simple and presumably quite frequent and robust. If *can* and *left* were now assigned to different categories, INFL and V respectively, what sort of inversion rule would be triggered? Notice first that the old forms could be perpetuated quite simply under the new grammar if the inversion rule were formulated with a disjunction (27).

27   Permute: NP $\left\{ \begin{array}{c} \text{INFL} \\ \text{V} \end{array} \right\}$

With such a rule there would be no discrepancy between input and output for the inversion data. However, we know that there was a

---

[11]For this kind of information about the history of English one relies heavily on handbooks such as the *Oxford English Dictionary* and Visser's monumental *An Historical Syntax of the English Language*.

discrepancy, which means that (27) was not triggered despite available data. This suggests quite strongly that (27) is not an attainable rule of grammar, that disjunctions of this sort do not exist. If the theory of grammar does not have this effect, it is hard to see why the change should have taken place and why there should have been a discrepancy between input and output; that discrepancy **requires** an explanation. Again we learn something about the limits to triggering experiences, about the nature of UG, and about the interaction between the two entities, and we learn by carefully distinguishing a child's input and output. Perhaps ironically, we can learn about a child's input from earlier stages of a language, despite limited records.

The sixteenth-century child heard (25b) and (26b), but such types were clearly not robust enough to trigger Negative and Inversion rules applying to Modal **and** to V. We know this because in fact such sentences dropped out of English and ceased to occur in the texts. Alongside these forms, the child also heard forms with *do: John did not go, did John go?* These were consistent both with grammar (22) (with Negative and Inversion applying to V, *did* being a V at this stage) and with grammar (23) (with the rules applying to Modal, *did* now being a Modal); (25b) and (26b) were consistent with grammar (22) and not (23) (assuming for all stages Negative and Inversion rules applying to only one category, either V or Modal). Therefore, as grammar (23) developed, so the *do* forms continued to occur; (25b) and (26b) did not serve to trigger any elaboration in the grammar that might have permitted them to survive alongside the *do* forms, and they dropped out of the language. I have not offered an explanation for this particular re-analysis, but I have used it only to illustrate how to discover when a re-analysis occurs.

Grammatical re-analyses meet strict conditions: they must lead to a grammar fulfilling the restrictive requirements imposed by UG; they must constitute the simplest attainable grammar for a child exposed to the new linguistic environment; they must yield an output close to that of the earlier grammars. For any re-analysis, these requirements impose narrow restrictions on the available options.

A grammar is not an object floating smoothly through time and space but a contingent object that arises afresh in each individual. Sometime one individual's grammar may differ significantly from that of her parents; that constitutes a new parameter setting. Assuming that changes over time can inform us about the limits to grammars, the study of diachronic change can show how idiosyncratic properties may be added to a grammar without affecting its internal structure, and what it takes to drive a grammar to re-analysis, with a parameter set differently. Examining historical re-analyses allows special insight into what kinds of trigger exeriences elicit grammatical re-analyses and what kinds are not robust enough to have that effect. The point at which re-analyses take place sheds light on the load that can be borne by derivational processes; the limits are manifested by the occurrence of the re-analyses; the re-analyses are manifested by the simultaneity of the relevant surface changes. In this way research on historical change informs work on a restrictive theory of grammar and is fully integrated with that general enterprise.

Not only does the mere occurrence of a re-analysis suggest things about the proper shape of the theory of grammar, but it also suggests the particular cluster of properties that the re-analysis encompasses. The theory of grammar includes parameters which are set on exposure to relevant experience. Fixing a parameter one particular way, say fixing a phrase structure rule as V → V NP, may have elaborate consequences for the form of somebody's knowledge, for the range of possible surface structures. We have seen several examples of this, and research aims to define the parameters as accurately as possible. If in historical change a parameter comes to be set differently, the precise definition given by the theory for that parameter will have implications, often far-reaching, for what exactly will change in the surface structures. Examining the cluster of properties encompassed by particular re-analyses often suggests things about how the theory should define its parameters. For example the changes discussed in this section suggest that the theory should allow grammars to use distinct categories of Verb and Modal; this is something denied by some people who argue that there is, even in the grammar of modern English speakers, no distinct category Modal.

It may be objected that this approach to language history accounts only for some changes and not others. Typically a theory works where it works and usually has no principled basis for not dealing with certain phenomena. In this particular domain one **must** allow a role for chance and non-grammatical factors; historical developments cannot and should not be totally predictable, for the reasons given earlier. Moreover, one cannot know in advance precisely which changes are due to chance factors and which are prompted by the theory of grammar. It is reasonable to suppose that changes involving the **loss** of certain sentence types, like those of (24), must be due to principled factors, because it is hard to imagine why a sentence type could cease to occur for reasons of stylistic force or foreign borrowing. Similarly, the change in the meaning of *like* is more likely to reflect a problem of attainability than a chance innovation for reasons of expressiveness. On the other hand, the gradual introduction of a new dislocation structure (*Mingus, I met him*) can plausibly be attributed to the need for strikingly new forms. But there are no firm guidelines.

The theory of grammar used here casts light on historical changes by explaining certain re-analyses; looking at the point at which re-analyses occur may suggest revisions to the theory of grammar. If a given theory explained change x and not change y, then x is a predictable re-analysis and y is due to non-grammatical factors. As in other domains, theories should be evaluated by the explanations that they offer. If a theory aimed only to explain historical changes, there would be a problem of indeterminacy here, and there would be no way to choose between theories that explained and attributed to chance different changes. In fact, the theory must meet many more demands, as shown earlier, and at this stage of research it is hard to imagine having to choose between a variety of theories all of which met the empirical demands that we are making. The problem, rather, is to find one adequate theory.[12]

4. Conclusion

There are several good reasons to view the growth of a child's linguistic capacity as a process whereby exposure to some linguistic environment sets certain parameters which are already available to the child independently of experience, presumably prescribed genetically in some fashion. This yields a system which enables the child to use and understand an indefinite range of novel utterances, as is usual. Taking this general perspective and adopting some specific ideas about the

---

[12]There is a potentially rich source of information about historical re-analyses in studies of what happens when a *pidgin* forms much of the linguistic environment of a child. A pidgin is an artificially simplified language used by colonial traders of different mother tongues. In those circumstances, the pidgin acts as a trigger and sets off in the child the development of a natural grammar in the usual way. The child naturalizes the pidgin, making it into a *creole*. That naturalization process, or creolization, typically involves many re-analyses occurring in quick succession form generation to generation, as a bona fide natural language emerges. It is often difficult to gather reliable data, but there are some interesting studies. For one example, see Sankoff and Laberge (1974).

types of genetically prescribed parameters enables us to understand several aspects of the way in which languages tend to change over the course of time. They do not change in arbitrary ways. There are certain kinds of changes which occur frequently: so, languages often adopt new word orders and main clauses are the initial locus of such innovations; morphological endings are formed and erode, often with syntactic consequences. There are also very specialized changes, like the change in meaning of English *like*. I have shown here how a certain theory of parameters would explain such diachronic phenomena. I have also shown how this parameter-setting view of language acquisition explains the fact that languages are constantly changing gradually and in piecemeal fashion and that they may also undergo a more radical restructuring from time to time. A process of gradual, piecemeal change punctuated by periodic radical changes is reminiscent of "punctuated equilibrium" models of evolutionary change (see Gould 1978, etc.) and of discussions in the context of Catastrophe Theory. What is interesting about linguistic change of this type is that it requires a particular kind of explanatory model, as I have tried to illustrate.

References

Allen, C.. "Reconsidering the History of *Like, J. Linguistics,* 22 (1986), 375–409.

Anderson, J.M., "A Note on Old English Impersonals,' *J. Linguistics,* 22 (1986), 167–177.

Baker, C.L., *Introduction to Generative-Transformational Syntax,* Prentice-Hall, Englewood Cliffs, NJ, 1978

Baker, C. L., "Syntactic Theory and the Projection Problem," *Linguistic Inquiry,* 10, No.4 (1979), 533–81.

Berwick, R.C., *The Acquisition of Syntactic Knowledge,* MIT Press, Cambridge, MA, 1985.

Changeux, J.-P., "Genetic Determinism and Epigenesis of the Neuronal Network: Is There a Biological Compromise Between Chomsky and Piaget?," in M. Piatelli-Palmarini, ed., *Language and Learning,* Routledge and Kegan Paul, London, 1980.

Changeux, J.-P., *L'homme neuronal,* Fayard, Paris, 1983.

Chomsky, N., *Aspects of the Theory of Syntax,* MIT Press, Cambridge, MA, 1965.

Chomsky, N., *Lectures on Government and Binding*, Foris, Dordrecht, Netherlands, 1981.

Elmer, W., *Diachronic Grammar: The History of Old and Middle English Subjectless Constructions*, Niemeyer, Tübingen, 1981.

Fischer, O. and F. van der Leek, "The Demise of the Old English Impersonal Construction," *J. Linguistics*, 19 (1983), 337–368.

Gould, S. J., *Ever Since Darwin: Reflections in Natural History*, Norton, New York, 1977.

Greenberg, J. H., "Some Universals of Grammar with Particular Reference to the Order of Meaningful Elements," in J. H. Greenberg, ed., *Universals of Language*, MIT Press, Cambridge, MA, 1966.

Hawkins, J. A., "Implicational Universals as Predictors of Word Order Change," *Language*, 55 (1979), No. 3, 618–648.

Hornstein, N. and D. W. Lightfoot, eds., *Explanation in Linguistics: The Logical Problem of Language Acquisition*, Longman, London, 1981.

Hyams, N., "The Pro-Drop Parameter in Child Grammars," *Proceedings of the West Coast Conference on Formal Linguistics*, 1983.

Jerne, N. K., "Antibodies and Learning: Selection versus Instruction," in G. C. Quarton, T. Melnechuk and F. O. Schmitt, eds., *The Neurosciences: A Study Program*, Rockefeller University Press, New York, 1967.

Jerne, N. K., "The Generative Grammar of the Immune System" *Science*, 229 (13 September 1985), 1057–1059.

Klima, E. and U. Bellugi, "Syntactic Regularities in the Speech of Children," in J. Lyons and R. Wales, eds., *Psycholinguistic Papers*, Edinburgh University Press, Edinburgh, 1966.

Lightfoot, D. W., *Principles of Diachronic Syntax*, Cambridge University Press, Cambridge, 1979.

Lightfoot, D. W., *The Language Lottery: Toward a Biology of Grammars*, MIT Press, Cambridge, MA, 1982.

Lightfoot, D. W., "The Child's Trigger Experience: Degree-Learnability," to appear *Behavioral and Brain Sciences*.

Matthei, E., "Children's Interpretation of Sentences Containing Reciprocals," in S. L.Tavakolian, ed., *Language Acquisition and Linguistic Theory*, MIT Press, Cambridge, MA, 1981.

Mazurkewich, I. and L. White, "The Acquisition of the Dative Alternation: Unlearning Overgeneralizations," *Cognition*, 16, No. 3 (1984), 261–283.

McNeill, D., "Developmental Linguistics," in F. Smith and G. A. Miller, eds., *The Genesis of Language: A Psycholinguistic Approach*, MIT Press, Cambridge, MA, 1966.

Monod, J. *Chance and Necessity*, Collins, London, 1972.

Newport, E. C., H. Gleitman and L. Gleitman, "Mother, I'd Rather Do It Myself: Some Effects and Non-effects of Maternal Speech Style," in Snow and Ferguson, eds., 1977.

Piattelli-Palmarini, M., "The Rise of Selective Theories: A Case Study and Some Lessons from Immunology," in W. Demopoulos and A. Marras, eds., *Language Learning and Concept Acquisition*, Ablex, Norwood NJ, 1986.

Randall, J., "Retreat Routes," Paper presented to the Boston University Conference on Languge Development, 1986.

Rizzi, L., "Comments on Chomsky's 'On the Representation of Form and Function'," in J. Mehler, E. C. T. Walker, and M. Garrett, eds., *Perspectives on Mental Representation*, Lawrence Erlbaum, Hillsdale, NJ, 1982.

Roeper, T., "Children's Syntax," manuscript, University of Massachusetts, Amherst, MA, 1979.

Sainsbury, R., "The 'Feature Positive Effect' and Simultaneous Discrimination Learning," *J. Exper. Child Psych.*, 11 (1971), 347–356.

Sainsbury, R., "Discrimination Learning Utilizing Positive or Negative Cues," *Can. J. Psych.*, 27, No. 1 (1973), 46–57.

Sankoff, G. and S. Laberge, "On the Acquisition of Native Speakers by a Language," in D. DeCamp and I. F. Hancock, eds., *Pidgins and Creoles: Current Trends and Prospects*, Georgetown University Press, Washington, D. C., 1974.

Snow, C., "Mothers' Speech Research: From Input to Interaction," in C. E. Snow and C. A. Ferguson, eds., 1977.

Snow, C. and C. A. Ferguson, eds., *Talking to Children: Language Input and Acquisition*, Cambridge University Press, Cambridge, 1977.

Warner, A., "Review Article on Lightfoot 1979," *J. Linguistics*, 19 (1983), 187–209.

Wells, G. et al., *Learning Through Interaction: The Study of Language Development*, Cambridge University Press, Cambridge, 1981.

Wexler, K. and P. W. Culicover, *Formal Principles of Language Acquisition*, MIT Press, Cambridge, MA, 1980.

INDEX

MATHEMATICAL
MODELING

Series editors:

William Lucas	Maynard Thompson
Department of Mathematics	Department of Mathematics
Claremont Graduate School	Indiana University
Claremont, CA 91711	Bloomington, IN 47405

Mathematical Modeling is a series of carefully selected books which present serious applications of mathematics for both the student and professional audience. The series aims to familiarize the user with new models and new methods and to demonstrate the art of constructing useful mathematical models of real-world phenomena.

We encourage preparation of manuscripts in LateX or AMS T_EX for delivery in camera-ready copy, which leads to rapid publication, or in electronic form for interfacing with laser printers or typesetters.

Proposals should be sent directly to the editors or to: Birkhäuser Boston, 675 Massachusetts Avenue, Suite 601, Cambridge, MA 02139.

MMO1 *Probability in Social Science*, Samuel Goldberg

MMO2 *Popularizing Mathematical Methods in China: Some Personal Experiences*, Hua Loo-Keng and Wang Yuan

MMO3 *Mathematical Modeling in Ecology*, Clark Jeffries

MMO4 *Newton to Aristotle: Toward a Theory of Models for Living Systems*, John Casti and Anders Karlqvist, eds.

MMO5 *Introduction to Queuing Theory, 2nd edition*, B.V. Gnedenko and I.N. Kovalenko (translated from Russian)